THE QUANTUM PHYSICS OF RELATIONSHIPS

Understanding Love and Connection Through the Lens of Quantum Mechanics

The Expanded Edition

Erica Fargo

SOTAPANNA PRESS

Published by Sotapanna Press.

This book describes ideas from physics and from relationship science and uses them to think about love. It is not a physics textbook, and it is not therapy. Nothing here substitutes for the care of a licensed clinician. If you are in a relationship that frightens you, please talk to someone who can help you in person.

First Edition published 2024. Expanded Edition published 2026.

ISBN-13: 979-8-1827-1218-7

Printed in the United States of America.

CONTENTS

PART FOUR

Change: What Moves

PART FIVE

Integration: What To Do With All This

A Note To The Reader Before We Begin

I fell into this book through a door I was not looking for.

Some years ago I was reading about the double slit experiment, badly and out of my depth, and I got to the part where the physicists discover that watching the electrons changes what the electrons do. And I put the book down, and I thought about a man I had loved for four years, and about how the version of him I got was so obviously a function of the version of me that showed up, and about how neither of us ever managed to see the other one plain.

That is a ridiculous thing to think while reading about electrons. I have not been able to stop thinking it since.

This book is what happened next. It is about the strangest ideas human beings have ever had about the nature of reality, and it is about love, and the argument I want to make is that these two subjects are not as far apart as they look.

Let me tell you what happened in physics, because it is one of the great stories.

For three hundred years the universe behaved. Things had positions. Causes preceded effects. If you knew enough, you could predict everything, and the only obstacle to knowing everything was effort. Then, in the space of about twenty-five years, a handful of mostly young people looked closely at very small things and found a world that did none of this.

They found that particles could be in combinations of states that have no equivalent in ordinary language. That the question you asked determined what kind of answer could exist. That two objects could be correlated in a way no distance could weaken and no signal could explain. That certain properties came in pairs, so that pinning one down smeared the other, not through clumsiness but through the nature of the things themselves. That a particle could pass through a wall it had no business passing through, and does so constantly, and this is why the sun is burning over your head right now.

They were horrified. Einstein spent the rest of his life objecting. Schrödinger said that if all this damned quantum jumping were to stay, he was sorry he had ever gotten involved.

And they were right to be horrified, because what they had actually discovered was that their categories had run out. Not their instruments. Their *categories*. The old words did not have the right shape to hold what was there.

Now. Consider the categories you were handed for love.

Love is a thing you find. People have a true nature you either discover or fail to discover. There is a fact of the matter about whether he loves you, and your job is to determine it. Certainty is the goal. Ambiguity is a problem. Distance weakens connection. Contradiction means somebody is lying.

Every single one of those is a classical assumption, and every one of them fails in exactly the way classical physics failed, for a reason that rhymes: they all assume that things have definite properties independent of how you engage with them.

I do not think your relationship is governed by quantum mechanics. I want to say that plainly once, here, so we can get on with it. Nobody's electrons are in love with anybody's electrons.

What I think is that the physicists built a set of tools for thinking about a world that would not behave, and that we are all living in a world that will not behave, and that they are not using the tools at the moment.

So I have borrowed them.

Five times in this book, and only five, you will hit a short section called **The Fine Print**. Those are the places where the true story is stranger and better than the mystical version people usually substitute for it, and I could not stand to let you have the substitute. The rest of the time I am going to let the metaphors run, and you can decide for yourself how far to take them. The universe does not need our help being astonishing, but it does not object to a little enthusiasm either.

There is also a chapter near the end where I take the whole thing apart and show you the places it does not hold. Read it whenever you like. It will not spoil anything.

One last thing. The couples in these pages are composites, stitched together from patterns that turn up constantly in the research and in ordinary life. None of them is one real person. You will meet four of them again and again.

Nadia and Wes, eleven years in, two children, and one argument wearing eleven years of costumes. Dev and Marcus, an ocean and five time zones apart, discovering that the hard part is not the missing. Priya and Tom, two years in, watching the fireworks stop and not knowing whether what is replacing them counts. And June, fifty-eight, widowed, dating again for the first time since the Carter administration and finding it both worse and better than advertised.

They are here because nobody has ever learned anything from an abstraction. We learn from specific people in specific trouble.

Come on. The sun is burning on a technicality and we have a lot to get through.

Introduction: The Two Mysteries

In 1927, in a Belgian hotel with better wine than it deserved, twenty-nine of the most capable minds on the planet gathered to argue about whether reality existed when nobody was checking.

The Fifth Solvay Conference on Photons and Electrons is remembered mostly for a single sustained argument between Albert Einstein and Niels Bohr. Einstein would arrive at breakfast with a thought experiment designed to prove that the new quantum theory was incomplete, that behind its probabilities there must be a deeper layer of definite facts. Bohr would spend the day dismantling it and present the demolition by dinner. This went on for the better part of a week. Einstein is supposed to have said that God does not play dice. Bohr is supposed to have replied that Einstein should stop telling God what to do. Both quotations have been polished by retelling, which is what happens to good lines.

What was actually at stake was not dice. It was something much stranger, and much more relevant to your marriage than it has any right to be.

Einstein believed that a particle has a position and a momentum at all times, that these are facts about the world, and that quantum mechanics simply fails to tell us what they are. The theory, in his view, was a very accurate ignorance. Bohr believed something more radical: that asking for the position and momentum of a particle simultaneously is not a hard question but a malformed one, like asking for the color of a smell. The properties are not sitting there waiting to be discovered. They come into definite existence in relation to a measurement, and different measurements bring different properties into existence, and some of those measurements are mutually exclusive.

Einstein wanted to know what was really there. Bohr said that "what is really there" was not a question physics could answer, and possibly not a question at all.

Here is why I am telling you this.

Ask almost anyone what the central problem of their relationship is and, underneath whatever they say about dishes or in-laws or money, you will usually find some version of Einstein's complaint. *I do not know what he really thinks. I cannot tell if she actually loves me or is just used to me. I need to know where this is going. Who is he when I am not in the room?*

Every one of those is a request for the hidden variable. For the definite fact underneath the appearance. For the thing that is true whether or not anyone is looking.

And every one of them runs into the same wall Einstein ran into, for reasons that are not identical to the physics but rhyme with it in a way that I have never been able to shake.

Two Mysteries That Rhyme

The twentieth century produced two great unsolved problems about the nature of the world.

The first is the measurement problem in quantum mechanics: why the mathematics describes a spread of possibilities but our experience delivers exactly one outcome, and what, exactly, makes the difference.

The second is the problem of other minds: how any of us can know what is happening inside another person, given that we have access only to their behavior, their words, and our own inferences, and given that all three are unreliable.

These are not the same problem. I want to say that clearly, because a lot of books in this neighborhood blur the two and then act as though the blurring were a discovery. They are different problems in different fields with different methods and almost certainly different solutions.

But they rhyme. Both involve a system whose interior is not directly accessible. Both involve an observer who cannot get information out without putting something in. Both produce the same characteristic frustration, which is the sensation of reaching for a fact and finding a relationship instead.

And here is the part that I find genuinely useful: physics got somewhere by giving up on the wrong question. Not by answering "what is really there when nobody looks," but by developing a rigorous account of what can be said, what can be predicted, what pairs of things can be known together, and what happens when a small isolated system comes into contact with a large messy environment. The theory works spectacularly. Your phone works because of it. The question Einstein wanted answered remains open, and physics has flourished anyway.

There is a lesson in that for the way you love someone.

What This Book Actually Contains

The chapters ahead are organized into four movements.

Measurement deals with what looking does. The observer effect, which is not what most people think it is and is more interesting than what they think it is. Complementarity, Bohr's idea that some questions exclude each other, which turns out to be the single most practically useful idea in this book. And decoherence, the mechanism by which a small private system loses its strangeness through contact with a large public world, which is the best account I know of what happens to couples over time.

Connection deals with what binds. Entanglement, the real thing rather than the greeting card version. Nonlocality and the surprisingly rigorous case for why distance does not do what you assume it does. And coherence, the condition in which many small components stay in step with one another and thereby accomplish things that no component could accomplish alone.

Indeterminacy deals with what cannot be pinned down. Superposition, and the underrated skill of holding two true things at once. The uncertainty principle, correctly stated, and its application to the oldest fight in the history of couples: how close is too close. And wave and particle duality, which is a genuinely good frame for the fact that love shows up as both a steady background condition and a series of discrete overwhelming events, and that people who want only one of those are going to be disappointed half the time.

Change deals with what moves. Tunneling, and why breakthroughs happen at the moment they seem least likely. Many worlds, and what to do with the life you did not choose. Entropy, which is not quantum mechanics at all, which I will admit up front, and which is nonetheless the most important physics in the book. Symmetry, and Emmy Noether's extraordinary result that the things you can count on are produced by the things that do not change, which turns out to be the best available account of trust. And phase transitions, including the reason that the path back into a good state is never the path by which you left it.

Then the reckoning: a chapter on where all of this falls apart, followed by a twelve-week practical program that requires no physics whatsoever.

A Word About Difficulty

You do not need any physics background. I mean that.

What you will need is a tolerance for being briefly confused, which is the ordinary condition of everyone who has ever thought seriously about this material, including the people who invented it.

I have used no mathematics. Not one equation. What I have done instead is tell you the real story rather than the cartoon, because the cartoons are not simplifications, they are errors, and errors make bad metaphors. The double slit experiment is not about particles knowing they are being watched. Entanglement is not instant communication. Superposition is not a thing being in two places. Every one of those, carried into a relationship, produces advice that will hurt you.

The true versions are better anyway. They are almost always better. That is the thing nobody tells you about physics.

The Promise

By the end of this book you should be able to do several specific things.

You should be able to notice the moment when your interpretation of your partner's behavior stops being a reading and starts being a construction, and to interrupt that process on purpose.

You should be able to identify which of your relationship goods are genuinely in tension with each other, so you can stop treating an inherent tradeoff as a personal failure.

You should be able to hold a contradiction, in yourself or in someone you love, for longer than the twelve seconds most of us can currently manage before we demand a resolution.

You should have a working theory of why closeness fades and a set of concrete moves for slowing it down.

And you should be able to tell the difference between a relationship problem that requires more information and one that requires a decision, which are almost never the same thing and are confused with each other constantly.

None of that requires you to believe that love is quantum. It requires only that you find the shape of the ideas useful, which I think you will, and that you be willing to sit with a little strangeness, which is good practice for the rest of it anyway.

PART ONE

Measurement: What Looking Does

Chapter One: The Observer Effect, Or Why Looking Changes What You Find

Nadia has been counting.

She has not told Wes this. She is not sure she would be able to explain it in a way that did not sound unhinged. But somewhere around the third year of their marriage she began keeping an informal tally of the evenings he came home and went straight to his phone, and by year eleven the tally has become something closer to a permanent background process, running underneath everything, requiring no effort at all.

Tonight he comes in at six forty. He sets his bag down. He says something about traffic on the 405. And then he takes out his phone.

What Nadia experiences in the next second is not a thought. It is faster than a thought. It arrives as a completed fact: *he would rather be anywhere than here.* By the time she is aware of having interpreted anything, the interpretation is already finished and has already been filed as evidence. Item four hundred and something.

What actually happened, if you had a camera on the room, is that a tired man looked at a screen for eleven seconds to see whether his mother had texted back about the biopsy results.

He does not mention the biopsy. He is not withholding; he simply has not decided how to say it and is waiting for a better moment, and the moment never quite arrives because Nadia's voice has gone flat, and he has learned over eleven years that when her voice goes flat, opening a serious subject will not go well.

Two people are now in a room together, and neither one of them is in the room the other one thinks they are in.

The Experiment That Would Not Behave

In 1801 Thomas Young settled what everyone assumed was a settled question. He let light pass through two narrow parallel slits and looked at what landed on the wall behind. If light were a stream of tiny bullets, you would expect two bright bands, one behind each slit. What Young got instead was a series of stripes: bright, dark, bright, dark, fanning out across the wall.

That pattern is the signature of waves. Where the crest of a wave coming through the left slit meets the crest of a wave coming through the right slit, they add and you get brightness. Where a crest meets a trough, they cancel and you get darkness. You can see the same thing in a harbor when two wakes cross. Light, Young concluded, is a wave. For a century that was the end of the story.

Then the twentieth century happened, and the experiment was run with electrons.

Electrons are not supposed to be waves. An electron is a particle with a mass and a charge; you can count them one at a time; they arrive at a detector as discrete clicks. And yet when you fire electrons at a screen with two slits, the pattern that builds up on the far side is the striped interference pattern. Waves.

Here is the part that made people stop sleeping properly. You can turn the beam down so low that only one electron is in the apparatus at any given moment. Fire one. Wait. Fire another. Each one arrives as a single dot, in one place, a particle in every observable respect. But let the dots accumulate over hours and they assemble themselves into the stripes. Each individual electron, arriving alone with nothing to interfere with, has somehow landed in a way that respects a pattern created by both slits.

And then the part that made it famous. Put a detector at the slits, so you can record which slit each electron went through. Change nothing else. Fire the electrons exactly as before.

The stripes vanish. You get two plain bands, the bullets pattern, the classical result.

Stop recording which slit. The stripes come back.

The Fine Print

The one thing worth knowing, because it makes the story better rather than worse: the electron is not shy. It has no opinion about your attention. Set up a which-path detector, record the data, delete the drive before any human reads it, and the stripes are still gone.

What matters is not whether anyone looked. It is whether the universe *could* have looked. The moment the world holds a record of which way the electron went, the magic stops.

Sit with that for a second, because it is stranger than the version you were told. It is not that consciousness collapses reality. It is that reality keeps receipts, and the receipts are enough.

Which, as it happens, is also true of marriages.

What Measurement Costs

The version of the observer effect that people usually cite is Heisenberg's original explanation from 1927, sometimes called the microscope argument: to see where an electron is, you have to bounce a photon off it, and that photon kicks the electron and changes its momentum. Look harder, kick harder.

Heisenberg later abandoned this as the explanation of the uncertainty principle, and modern physics regards it as not quite right for that purpose. But as a description of a real physical effect, measurement disturbance is entirely genuine. Getting information out of a system requires putting energy into it. There is a price, and the price is paid by the thing you are studying.

Now hold that next to Nadia counting.

Her surveillance is not free. It costs Wes something. He is being observed, continuously, by an instrument calibrated to detect withdrawal, and the instrument is sensitive, and it has been running for eight years. And a person under that kind of observation does what anything under continuous measurement does: he changes. He becomes careful. He rehearses. He avoids the topics that produce the flat voice, which is most of the topics that matter, which means he is now, in fact, less present than he would otherwise be.

The measurement produced the result the measurement was looking for. Not because Nadia is unfair. Because measurement is not free.

The Research That Backs This Up

I want to be careful here, because this is exactly the point at which a book like this usually starts waving its hands. So let me give you the actual literature.

Behavioral confirmation. In a well-known 1977 study, Mark Snyder, Elizabeth Tanke, and Ellen Berscheid had men have telephone conversations with women they had never met. The men were shown a photograph beforehand, supposedly of their conversation partner, though in fact the photographs were assigned at random and depicted either a conventionally attractive or unattractive woman. The men who believed they were talking to an attractive woman behaved with more warmth. And the women on the other end, who knew nothing about any photograph, responded with more warmth and animation. Independent judges listening only to the women's half of the conversation, with no knowledge of the manipulation, rated those women as friendlier and more likable. The expectation, held privately by one person, had reached across and changed the behavior of another. Nobody had to say anything.

Attributions in marriage. This is the most directly relevant body of work, and it is robust. Beginning in the 1980s, Frank Fincham and Thomas Bradbury documented what they called a distress-maintaining attributional style. When a happy spouse experiences a negative act by their partner, they tend to explain it situationally and narrowly: *he is quiet tonight because work has been brutal, and it will pass.* When a distressed spouse experiences the identical act, they explain it dispositionally and broadly: *he is quiet* *because he is selfish, and he has always been selfish, and he will be selfish about the next thing too.* The behavior is the same. The explanation is not.

The critical finding is that the direction of causation runs both ways. Longitudinal studies found that attributional style at one point in time predicted marital satisfaction later, controlling for satisfaction at baseline. The explanations are not simply a readout of how the marriage is going. They are among the things that determine how the marriage will go.

Positive illusions. Sandra Murray, John Holmes, and Dale Griffin published a series of studies in the 1990s asking whether it is better to see your partner accurately or generously. The answer was more interesting than either camp expected. Partners who idealized each other, meaning they rated each other more positively than those partners rated themselves, were happier, and stayed happier over time, and their relationships were less likely to end. More strikingly, over the course of a longitudinal study, people tended to drift toward the idealized version. Being seen well made people better. Murray and Holmes called this the phenomenon of prophecies coming true.

The caveat that keeps this from being a license for delusion: the effect operates on trait-level generosity, not on the denial of specific facts. Believing your partner is kinder than they believe themselves to be is protective. Believing your partner did not do the thing they demonstrably did is not idealization, it is something else, and it does not help anybody.

The one you should be skeptical about. You may have heard of the Pygmalion effect, from Robert Rosenthal and Lenore Jacobson's 1968 study in which teachers were told certain randomly selected students were poised for a growth spurt, and those students subsequently showed larger IQ gains. It is a famous study, and I am not going to pretend the replication record is clean. Effects have been inconsistent and are generally smaller than the original suggested. I mention it because you will encounter it, and because a book that only tells you about the studies that support its thesis is a book that should not be trusted.

The expectancy literature in close relationships, however, is stronger than the classroom literature. Partners have far more contact, far more emotional stake, and far more channels through which expectation can leak into behavior.

What The Metaphor Gets Right, And Where It Stops

The parallel is genuine on one specific point: in both cases, there is no such thing as a purely passive observation. Getting information requires interaction, and interaction alters what you were measuring.

The parallel fails on several others, and you should know them.

In physics, the disturbance is symmetric, mechanical, and quantifiable. In relationships, the disturbance runs through interpretation, memory, and emotion, all of which are messy and none of which are described by an equation. Physics has no equivalent of Nadia's eight years of accumulated tally.

In physics, the system being measured has no model of the measuring device. Wes has a very detailed model of Nadia's measuring device. He has been reverse engineering it for a decade. That feedback loop, the observed system modeling the observer and adjusting, has no analogue in quantum mechanics at all, and it is arguably the single most important dynamic in a long relationship.

And in physics, there is no question of fairness. Here there certainly is. Which brings us to the part that most books in this genre skip.

The Part About Power

If observation shapes reality, then observation is a form of power, and the ethics of it are not optional.

There is a version of everything I have just told you that becomes an instruction to gaslight yourself. *See him generously* can slide into *do not trust what you noticed. Your expectations create your reality* can slide into *the problem is your perception, not his conduct.* I have seen this happen to people. I have seen women in particular be handed the observer effect as a reason to disbelieve their own pattern recognition, and it is a small catastrophe every time.

So let me draw the line as sharply as I can.

The observer effect applies to interpretation, not to events. Whether Wes looked at his phone is an event. Whether that means he would rather be elsewhere is an interpretation. The first is not up for negotiation. The second is where all the room is.

Generous interpretation is a strategy for ambiguity, not for clarity. When the data are genuinely underdetermined, which covers the enormous majority of daily relational life, generosity is both kinder and more accurate, because the charitable reading is usually the true one. When the data are not underdetermined, when there is a pattern, when it has been named and it has not changed, generosity is no longer interpretation. It is denial, and it has a different name and different consequences.

A measurement you are afraid to disclose is surveillance. This is the cleanest test I know. If you could tell your partner what you are tracking and how, and you would be willing to hear what they track about you, you are in a relationship. If the tally must stay secret because naming it would expose how much it has grown, the tally has become something else. Nadia's count has been secret for eight years. That secrecy, not the count itself, is the actual problem.

Practice: The Two-Column Method

This is the only exercise in this chapter, and it is worth doing properly rather than doing five things badly.

Take a sheet of paper and draw a line down the middle. On the left, write what a camera would have recorded. Only that. Physical events, spoken words, timestamps. No adjectives that require a mind reader. "He looked at his phone for eleven seconds" is camera data.

"He looked at his phone dismissively" is not, because dismissiveness is not visible; it is inferred.

On the right, write what you concluded.

Most people find the left column is three lines long and the right column runs off the page. That ratio is the finding. You are not observing your partner. You are observing a very small amount of your partner and then generating a great deal of content about them, at speed, using a model built years ago from data that may no longer apply.

Now the second half, which is the part people skip. For your central right-column conclusion, write down three alternative explanations that would produce identical left-column data. Not nicer explanations. Different ones. The bar is that a reasonable stranger watching the footage could not distinguish between them.

For Nadia, that might be: he is avoiding me. He is exhausted. He is dealing with something he has not found a way to say.

Only one of those was true. She had never generated the third one, because her instrument was not built to detect it.

What usually goes wrong with this exercise: people write alternatives they do not actually consider plausible, as a compliance move, and then feel superior for having done the assignment. If none of your three alternatives makes you feel even slightly uncertain, you have not done it yet. Uncertainty is the deliverable.

And then the part that is not on paper: you have to ask. The two-column method generates hypotheses. It cannot select between them. The entire exercise is worthless if it ends with you privately deciding which alternative is correct, because that is just the original problem with extra steps. It ends with a question asked out loud, in a neutral voice, at a time when nobody is already angry. "You went quiet when you came in and I noticed I had a story about it. Can I check the story with you?"

Nadia And Wes, Continued

She asked him on a Thursday, badly, with more edge in her voice than she intended, which is how these things usually go and is fine.

He told her about his mother. He told her he had been carrying it for nine days and had been waiting for a moment when the room felt open, and that the room had not felt open in some time, and that he was aware this was partly his own doing.

They did not resolve eleven years of this in one evening. That is not how any of this works, and any book that suggests otherwise is lying to you for commercial reasons.

But something in the apparatus changed. Nadia had spent eight years running a measurement that could only ever return one of two values, present or absent, and had never once considered that the variable she was measuring might not be the variable that mattered. When she finally asked a genuinely open question, she got back information her instrument had never been capable of detecting.

That is the whole chapter. Your instrument determines your data. If you have been getting the same answer for years, consider that you may have been asking a question with only one possible answer.

Which is, as it happens, exactly the problem Niels Bohr spent his life on. That is the next chapter.

Chapter Two: Complementarity, Or How The Question Decides Which Answer Can Exist

Priya has a specific complaint about Tom, and it is the most common complaint in the recorded history of couples, which does not make it less real.

She comes home upset about her manager. She describes the meeting. Tom listens for approximately forty seconds and then says, with real warmth and genuine intent to help, "Okay, so what if you just went to Anita directly?"

And Priya feels something close inside her chest, and she cannot explain why, and when she tries to explain why it comes out as an accusation, and then they are having a different fight than the one about her manager.

Tom's suggestion was good. That is the maddening part. Going to Anita directly was, in fact, the correct move, and Priya did it two weeks later and it worked. Tom was right and Tom was helpful and Tom made it worse.

There is a standard explanation for this, which you have heard, and which is wrong. The standard explanation is that women want to be heard and men want to fix things. This is wrong on the gender claim, which does not survive contact with actual data, and wrong on the deeper structure, which is not about what people want but about what is possible.

The real answer is that Priya was running one kind of measurement and Tom performed a different one, and the two are mutually exclusive, and no amount of goodwill makes them compatible.

Niels Bohr had a word for this.

The Idea Bohr Cared About Most

Bohr is remembered for the atom with the little orbits, which he would have found irritating, since he abandoned it fairly quickly and considered his real contribution to be something much more general.

That contribution is complementarity, announced at Como in 1927 and refined for the rest of his life. The claim is this: for certain systems, there exist descriptions that are each valid, each necessary for a full account, and mutually exclusive in application. You cannot deploy both at once. Not because your equipment is bad. Because the experimental arrangement required for one excludes the arrangement required for the other.

The canonical case is wave and particle. Set up an apparatus that reveals interference and you have an unambiguous demonstration of wave behavior. Set up an apparatus that determines which slit the electron traversed and you have an unambiguous demonstration of particle behavior. Both are correct. Neither is the deeper truth behind the other. And the apparatus that gives you one is physically incapable of giving you the other, because the arrangements are incompatible at the level of hardware.

Bohr's radical move was to say that this is not a limitation on our knowledge of the electron. It is a statement about what "the electron" means. There is no context-free electron sitting behind the experiments with both properties simultaneously definite, waiting for a cleverer physicist. The properties are relational. They exist in the coupling of system and apparatus. Change the apparatus and you have not gotten a worse view of the same thing; you have brought a different thing into existence as a definite fact.

Einstein hated this. Many physicists still do, and there are serious alternative interpretations that avoid it. But as a description of how quantum experiments actually behave, complementarity has never been contradicted by a measurement, and modern experiments with which-path information have made it sharper rather than weaker.

Two Instruments, One Room

Return to Priya and Tom.

When Priya describes her manager, she is not transmitting a problem. She is running what psychologists call a responsiveness check. Harry Reis and Phillip Shaver, in an influential 1988 model of intimacy, described the process as a loop: one person discloses something that matters, the other responds, and the discloser then evaluates whether the response conveyed three specific things. Did you understand me. Do you value what you understood. Do you care.

Notice that a solution answers none of those three. A solution answers a fourth question that was not asked.

Tom, meanwhile, is running a competence measurement, and it is not a selfish one. He loves her, she is in pain, and he has been raised in a world that treats the alleviation of a loved one's pain as the entire content of love. Offering the Anita suggestion is, in his internal accounting, an act of care that costs him effort and delivers value.

Here is the incompatibility. Reis and Shaver's responsiveness loop requires the responder to stay inside the discloser's frame long enough to reflect it back. Problem-solving requires the responder to step outside the frame, evaluate it from a distance, and re-describe the situation in terms of actions available. Those are opposite postures. Inside and outside. You cannot occupy both positions at the same moment for the same content, any more than one apparatus can both preserve interference and record which path.

And the cost is not symmetric, which is the cruel part. If Tom does responsiveness first, Priya can receive the Anita suggestion twenty minutes later and be grateful for it. If Tom does problem-solving first, the responsiveness channel closes and cannot be reopened in that conversation, because Priya now has data: her distress was treated as a malfunction to be corrected. Nothing Tom says afterward can un-collapse that.

Order matters. In quantum mechanics this has a name too. Operators that do not commute give different results depending on which you apply first, and that noncommutativity is the mathematical core of the whole business. Measure position then momentum and you do not get what you get measuring momentum then position.

Conversations do not commute either.

The Larger Pattern

Once you have the shape, you start finding it everywhere, and most of the intractable arguments in a long relationship turn out to be complementarity problems being treated as truth problems.

Being known and being desired. There is a real tension, described most clearly by Esther Perel in her work on eroticism in long partnerships, between the conditions that produce security and the conditions that produce desire. Security wants predictability, transparency, and the absence of distance. Desire, empirically, seems to require some novelty, some otherness, some space across which to reach. This is not a claim that you must choose one forever. It is a claim that the arrangement which maximizes one tends to suppress the other at any given moment, and that couples who understand this stop diagnosing a normal oscillation as a defect.

Honesty and kindness. Every couple discovers eventually that full transparency and maximal gentleness are not simultaneously achievable, and that people who insist otherwise are usually being neither. The question is not which one is the real virtue. The question is which arrangement this particular moment calls for, and whether you can be explicit with each other about which one you are running.

Autonomy and coordination. Chapter Eight is entirely about this pair, because it maps onto the uncertainty principle so cleanly that it deserves the space.

Grieving and rebuilding. June discovered this one the hard way. Eighteen months after her husband died, her sister kept asking when she was going to start living again, and June kept feeling that the question was a small violence, and she could not say why. It was because remembering him fully and constructing a life without him are complementary arrangements. She could do either. She could not do both in the same hour, and being asked to do both simultaneously felt like being asked to be a wave and a particle on demand.

What To Actually Do

The practical move is smaller than you would expect and works better than it has any right to.

Name the arrangement before you run it. Not in a therapeutic voice. In three words, in the doorway. "I need to vent." "I want ideas." "I don't know yet, just be near me."

Tom and Priya adopted a version of this that took about four seconds and eliminated a category of fight they had been having for two years. She learned to open with "this is a listening one." He learned that if she had not said, he could ask, and that asking was not a failure of intuition but a piece of equipment configuration.

The objection people raise is that having to ask ruins it. If he really knew me, he would know. I understand the feeling and I think it is a mistake. The demand that your partner correctly infer which experiment you are running, from ambiguous data, under time pressure, while managing their own state, is a demand for mind reading. It has a very low success rate and a very high cost of failure, and it is completely unnecessary, because you have language.

Do not ask which mode is right. They are both right. The question that dissolves the fight is not "should Tom have listened or solved" but "which arrangement was this conversation, and did we agree on it." Almost every complementarity fight is two people each defending the validity of their mode, which is doomed, because both modes are valid and neither person is going to concede a point that is actually true.

Sequence deliberately. When both modes are genuinely needed, and they often are, responsiveness goes first. Always. This is not a courtesy; it is a structural fact about what closes what. You can go from being heard to solving a problem. You cannot reliably go the other way.

Watch for the mode you never run. Most people have a default arrangement they deploy so consistently that they have forgotten it is a choice. Tom's is competence. Nadia's is threat detection. June's, for most of her marriage, was harmony maintenance, which is why nobody in that house discovered until much later that her husband had been unhappy for a decade. Your default is invisible to you and completely obvious to your partner. You can just ask them what it is. They will know.

An Exercise Worth Doing

For one week, before any conversation that has emotional weight, one of you says a single sentence naming what you need from the next ten minutes. That is the whole exercise. No journaling.

The three most useful labels I have found:

Witness. I need you to hear this and not do anything with it.

Counsel. I want your actual opinion, including the parts I will not like.

Company. Nothing needs to be discussed. I want to not be alone.

What usually goes wrong: the person who habitually solves will name "witness" and then solve anyway at minute six, because the urge is strong and the silence feels like failure. If that happens, the correct response is not a fight. It is one word: "witness." Said flatly, without accusation. It works because you already agreed on the vocabulary, and correcting an instrument setting is a much smaller event than accusing someone of not caring.

The other failure is subtler. Some people use "witness" to make themselves unchallengeable, declaring listening mode as a way to fence off a subject from any input forever. If every conversation is witness mode, that is not complementarity. That is a wall with a physics vocabulary.

Why This Matters More Than It Sounds Like It Does

Underneath the practical mechanics there is a philosophical shift, and it is the one Bohr spent thirty years trying to get people to make.

The classical instinct is that there is one true description of a situation and the disagreement exists because someone has it wrong. Under that instinct, every conflict is a search for the error, which means every conflict has a loser, which means every conflict is a threat.

The complementary instinct is that there can be two accounts, each complete within its arrangement, neither reducible to the other, both required for the full picture. Under that instinct, a disagreement can be a sign that you are looking at something rich enough to require two descriptions.

That is not relativism and it is not an excuse. Some claims are simply false, and complementarity does not rescue them. Bohr was not saying that any two contradictory statements about an electron are equally good; he was saying something precise about specific pairs of well-defined properties under specific conditions.

The relationship version needs the same discipline. "He never helps" and "he helps constantly" are not complementary descriptions. One of them is closer to true and you can go count. But "I need more closeness" and "I need more room" can both be true, in the same house, on the same evening, held by two people who love each other, and no amount of arguing will make one of them the error.

Knowing which kind of disagreement you are in is most of the skill.

Chapter Three: Decoherence, Or How A Private World Leaks Into The Public One

The single most common thing people say about a relationship that is dying is that nothing happened.

There was no affair. Nobody was cruel. There is no scene to point to, no line that was crossed, no moment you could film. There is only the observation, arriving one ordinary evening with the force of a diagnosis, that the two of you have become people who live near each other.

This chapter is about the mechanism.

I think it is the most important chapter in the book, and I think it is the one the original edition of this book got most wrong, because I wrote that edition without understanding decoherence properly and used the word "collapse" the way everyone does, which is to say incorrectly.

The Question Nobody Could Answer Until 1970

For fifty years quantum mechanics had a hole in it that everybody could see and nobody could fill.

The mathematics describes systems evolving smoothly through combinations of possibilities. An electron does not have a definite path; it has an amplitude spread across all of them, and those amplitudes interfere with each other, and the interference is real and measurable. Fine. But nobody has ever seen a coffee cup in a combination of on-the-desk and on-the-floor. Large objects have definite properties. Small ones do not. Where is the boundary, and what enforces it?

The standard answer, taught for decades, was that measurement collapses the wave function. Something about the act of observation converts a spread of possibilities into a single actuality. This was never a mechanism. It was a placeholder with an air of authority, and it invited exactly the mysticism that has attached itself to quantum physics ever since, because if consciousness collapses reality then the door is open to anything.

Beginning around 1970, with H. Dieter Zeh's work and then substantially through Wojciech Zurek in the 1980s, a real mechanism arrived, and it turned out to require no observers at all.

Here it is. Nothing is ever truly isolated. A dust grain in the deepest vacuum available is still being struck by photons from the cosmic microwave background, roughly a hundred billion times per second. Every one of those collisions leaves the photon slightly correlated with the dust grain's position. The information about where the grain is does not stay with the grain; it radiates outward into the environment, encoded in scattered light, in air molecules, in thermal vibrations.

And once that information exists out in the world, the interference terms that make quantum behavior quantum are gone. Not hidden. Gone from any measurement you could perform on the system itself. The possibilities stop being able to interfere with each other, because they are now distinguishable by their environmental fingerprints, and the system's behavior becomes indistinguishable from an ordinary classical thing with a definite property.

That is decoherence. Quantum strangeness is not destroyed by observation. It leaks away through contact.

The numbers are brutal. A large molecule in a good laboratory vacuum can maintain coherence for a fraction of a second. A dust grain, ten microns across, in intergalactic space with only starlight hitting it, decoheres in something like a millionth of a second. In a warm room, a comparable object loses coherence in something on the order of ten to the minus thirty-one seconds. This is why you have never seen a coffee cup in superposition and never will. The universe is not hiding quantum behavior from you. It is being scrubbed out, continuously, by contact, faster than you could ever hope to look.

The Fine Print

Decoherence explains why the world looks solid and ordinary. It does not fully explain why you experience one outcome rather than another, and that leftover puzzle is where the interpretations start fighting. Chapter Eleven has the popcorn.

But here is the part that is settled, and that I find genuinely thrilling:

Coherence is not the natural state of things. It is an achievement, maintained against constant erosion, and it requires active work to survive.

Every quantum computer on Earth is an argument for that sentence. To keep a handful of qubits in step long enough to compute anything, engineers chill them to a hundredth of a degree above absolute zero, hang them in vacuum, wrap them in shielding, and then run error correction that costs more than the calculation.

The coherence is not preserved by leaving it alone. It is preserved by relentless, expensive, unglamorous intervention against a universe that is always taking it back.

Now let us talk about your marriage.

The Private World

In 1964 two sociologists, Peter Berger and Hansfried Kellner, published a short paper called "Marriage and the Construction of Reality" that has aged better than most of what has been written about couples since.

Their argument was that a marriage is not primarily an emotional arrangement or a legal one. It is a world-building project. Two people arrive from separate social universes, each carrying a full set of assumptions about what things mean, what is funny, what is rude, how a Sunday goes, what counts as an emergency. In conversation, continuously and almost entirely unconsciously, they construct a third universe that belongs to neither of them and is inhabited by both.

The mechanism, they said, is talk. Ordinary, redundant, apparently pointless talk. The recounting of the day. The rehashing of an event you both attended. The relentless narration of trivialities. Through that talk, a shared account of reality is built and then continuously re-ratified, until the couple has a private nomos: their own history, their own vocabulary, their own settled interpretations of relatives and friends and the past.

You know this world from the inside. It has jokes that cannot be explained to anyone. It has a word that means something specific and would be meaningless to a stranger. It has a version of your worst year that the two of you tell together, with parts, and which is not identical to the version either of you would tell alone. Berger and Kellner noticed something remarkable, which is that the couple's own history gets rewritten by this process, so that each partner's past is gradually reinterpreted to make sense as leading toward the other.

That private world is a coherent state. Fragile, valuable, and not the natural condition of two adults sharing a mortgage.

How It Leaks

Decoherence, in the relational sense, is what happens when the private world comes into continuous contact with the public one and its distinctness bleeds out.

The mechanism is contact, and it is almost never dramatic.

The environment is other people. Your mother's opinion of your husband. Your friends' framework for what your marriage is. Your colleague's assumptions. Each contact carries a little of your private interpretation out and brings a little of the public interpretation in. This is why couples who complain constantly about each other to third parties tend to get worse rather than better:

the interpretation that gets rehearsed with outsiders becomes the operative one, and it is the public version, which is always flatter and less generous than the private one.

The environment is attention. Nadia and Wes have not had a conversation without a device present in three years. Not a serious one; a *conversation*. The device is not evil. It is a channel through which the outside world enters the room continuously, and every entry is a small loss of the enclosure inside which a private reality can be built. The talk that Berger and Kellner identified as the engine of world-building is precisely the low-value, redundant, unhurried talk that a phone eliminates first, because that kind of talk cannot compete with an infinite feed for attention and was never supposed to have to.

The environment is logistics. This is the big one, and it is why children are the classic decoherence event. Nothing about children destroys love. What children do is convert nearly every available conversational slot into operational coordination. Pickup. Dosage. The form. The other form. Coordination talk is not world-building talk; it is transactional, it is about the public world, and it fills every gap. A couple can go five years exchanging four thousand sentences a week, of which perhaps eleven are about anything that constructs a shared interior.

The environment is time itself. Even without a single stressor, the ratio drifts. The proportion of your communication that is instrumental rises. The private vocabulary stops acquiring new terms. The story of your worst year gets told less often and eventually only at funerals.

And here is the property that makes this so hard to see while it is happening, and which the physics captures precisely: **decoherence is gradual, continuous, and produces no event.** There is no moment. There is no line. There is only a system that was in a delicate correlated state slowly becoming a system that is not, through billions of individually negligible contacts, none of which was the cause.

Which is exactly what people mean when they say nothing happened.

Why "We Grew Apart" Is A Physics Statement

I want to defend a phrase that gets treated as an evasion.

When people say they grew apart, therapists and friends often hear a euphemism concealing a real cause. Sometimes it is. But often it is a completely accurate description of a decoherence process, and the reason it sounds evasive is that our vocabulary for relationship failure is built entirely around events. Betrayals, revelations, fights. We have rich language for catastrophe and almost none for erosion.

This has consequences. Couples in a slow decoherence often wait for something to happen, because they believe that a relationship in trouble should present with a symptom. Nothing presents. The absence of a crisis is read as the absence of a problem, right up until one of them says the sentence about living near each other, at which point five years of accumulated loss arrives at once and feels like a betrayal even though nobody did anything.

If you take one thing from this chapter: **the absence of conflict is not evidence of coherence.** Some of the healthiest couples fight regularly. Some of the deadest ones have not raised their voices in a decade. Conflict is a signal of contact. Contact is the thing you are actually trying to preserve.

Error Correction

Here is where the physics stops being a diagnosis and starts being an instruction.

Nobody builds a quantum computer by trying to eliminate decoherence, because that is impossible. You cannot cut a system off from the universe. What you do instead is two things: reduce the coupling where you can, and run continuous active correction that restores coherence faster than the environment degrades it.

Both have relational versions, and I think this is the most practical framework in the book.

Reduce the coupling. This means creating enclosures. Not romantic getaways, which are expensive, rare, and asked to carry more weight than any single event can. Small, regular, cheap enclosures with a real boundary.

The boundary is what matters, and it has to be physical, because willpower does not survive contact with a notification. Devices in another room, not face down on the table. A specific window that is defended against logistics: Nadia and Wes eventually adopted a rule that the first twenty minutes after the children were down could not contain a single sentence about scheduling, and they were both surprised by how much silence there was at first, and by what came into it after nine or ten days.

Run error correction. This is the more interesting half, because it is counterintuitive.

John Gottman's most cited finding is about the ratio of positive to negative interactions in stable couples, roughly five to one during conflict. The number is often quoted without its context and the underlying predictive models have drawn real methodological criticism, particularly about overfitting to small samples, which I mention because you should know it. But the qualitative core has held up across many labs: stable couples are not couples that avoid rupture. They are couples that repair it quickly and often.

Gottman's term for the mechanism is the repair attempt: any move, however clumsy, that interrupts an escalating spiral. A joke. A hand on an arm. Saying "wait, can we start over." What distinguishes durable couples is not the elegance of their repair attempts, which are often terrible, but that the attempts get *received*. A bad repair attempt that lands beats a beautiful one that is refused.

That is error correction. It does not prevent the error. It catches it and restores the state before the degradation accumulates. And, like error correction in a quantum computer, it must be continuous, it must be cheap enough to run constantly, and it consumes more resources than the thing it protects, which is not a design flaw but the price of the whole enterprise.

The frequency matters more than the magnitude. One weekend away per year does not compensate for fifty-one weeks of decoherence, for the same reason that a single heroic correction cannot restore a state that has already leaked out. The environment is acting on you every hour. Your correction has to run on a comparable schedule or it loses.

An Enclosure Protocol

Here is a concrete version, which Dev and Marcus developed for a long-distance relationship and which works equally well for people who share a bed.

Twenty-five minutes. Same days each week, and they must be actual days on an actual calendar. Monday and Thursday work well; pick whatever holds.

No devices in the room. Not silenced. Not present.

Three prohibited categories: scheduling, money, and any problem that has a to-do item attached. Those subjects are not banned from your life; they are banned from this window, because they will otherwise consume all of it, since they always feel more urgent than world-building and they are always wrong about that.

What goes in the window instead: recounting. Berger and Kellner's engine. Tell each other things that do not need to be told. Something odd you noticed. A memory that surfaced for no reason. A thing you have been thinking about that has no conclusion and no action item. The redundancy is the point. You are not exchanging information. You are re-ratifying a shared world.

What usually goes wrong: week two, one of you brings up the plumber. Week three you skip it because of a genuine emergency, and then again the next week because the streak is broken. Week five you conclude it was artificial and stop.

The artificiality objection is the one that kills this, so let me address it directly. Yes, it is artificial. So is brushing your teeth. The private world of a new couple builds itself effortlessly because the two of you have unlimited unstructured time and are fascinated by each other, which are environmental conditions, not character traits. When those conditions disappear, the process that they enabled does not continue on its own out of love. It has to be engineered. Refusing to engineer it because engineering is unromantic is how people end up single and confused about why.

What This Explains

A few things fall out of this frame that are otherwise puzzling.

Why couples reconnect on vacation and lose it within a week of returning. The vacation was an enclosure. You did not repair anything; you removed the environment temporarily. The coherence returns immediately, which is genuinely good news, because it means the state was recoverable. Then you go home and recouple to the environment and it drains at the usual rate.

Why long distance sometimes works better than expected. Dev and Marcus get eleven hours a week of undivided attention with nothing else in the room. Many cohabiting couples get less. Distance is a terrible way to run a relationship in most respects, and it turns out to be an accidental enclosure.

Why the empty nest is dangerous. For eighteen years the coordination talk was the entire channel. When the coordination ends, couples discover whether anything else was being built underneath, and a great many discover that nothing was, which is not a failure of love but a failure of maintenance.

Why grief decoheres a person. June's private world had two inhabitants and now has one, and a shared reality cannot be maintained unilaterally. Nobody else knows the vocabulary. Nobody else can re-ratify the story. This is why the loneliness of widowhood is not the same as the loneliness of being alone, and why the well-meaning suggestion that she get out more misses by a mile.

The good news, the thing decoherence gets right and pessimism gets wrong, is that a coherent state is not a fragile treasure you were lucky to have and have now lost forever. It is a dynamic condition, maintained by process, and processes can be restarted. Not easily. Not by accident. But the physics contains no rule that says a system, once decohered, cannot be brought back into a correlated state.

It says only that it will not happen by itself, and that leaving it alone guarantees the outcome you are afraid of.

PART TWO

Connection: What Binds

Chapter Four: Entanglement, Or The Correlation That Has No Messenger

Of all the words physics has lent to popular culture, entanglement has been abused the worst.

It appears on jewelry. It appears in wedding vows. It appears in a genre of book that suggests your love for someone is instantiated in correlated particles, that a sufficiently intense bond creates a physical channel, that when you thought of your sister at the exact moment her car went off the road, something quantum occurred.

Nothing quantum occurred. I want to clear this out completely, because underneath the nonsense there is an idea so much better than the nonsense that I resent the trade.

What Was Actually At Stake

In 1935 Einstein, along with Boris Podolsky and Nathan Rosen, published a paper intended to demonstrate that quantum mechanics could not be a complete description of reality. The argument ran roughly like this.

Prepare two particles together so that their properties are linked by a conservation law. Total spin, say, must be zero. Send them in opposite directions. Now measure one. Whatever you find, you instantly know what a measurement on the other one would give, because the two must sum to zero.

Einstein's point was that this creates a dilemma. Either the second particle had its property all along, in which case quantum mechanics is incomplete for failing to describe it, or the measurement on the first particle somehow instantaneously determined the state of the second one across an arbitrary distance, which Einstein called spooky action at a distance and considered obviously absurd.

He expected physicists to take the first option. Of course the particles had their properties all along. Of course there are hidden variables. This is a pair of gloves in two boxes: opening one and finding the left glove does not magically create a right glove in the other box, it merely informs you of a fact that was true since packing.

The glove argument is excellent. It is also, as we now know experimentally, wrong.

Bell's Knife

For nearly thirty years the EPR question was regarded as philosophy. Then in 1964 John Stewart Bell, an Irish physicist at CERN working on the problem in his spare time, did something nobody had thought possible: he turned it into an experiment.

Bell's insight was that the glove hypothesis, in fact any theory in which the particles carry predetermined values and no influence travels faster than light, makes a quantitative prediction. If you measure the two particles along a variety of different axes, not just the same one, and tabulate how often the results agree, any such local realistic theory imposes a hard ceiling on the correlations. Quantum mechanics predicts correlations that exceed that ceiling.

This is a remarkable thing. Bell found a number that separates two entire families of possible universes, and it can be measured on a bench.

The experiments were done. Stuart Freedman and John Clauser in 1972. Alain Aspect and his group in the early 1980s with switching that closed a critical loophole. Anton Zeilinger and others through the 1990s and 2000s. In 2015, three independent groups closed the remaining loopholes simultaneously in what are called loophole-free Bell tests. Clauser, Aspect, and Zeilinger shared the 2022 Nobel Prize for this body of work.

The ceiling is violated. Every time. By exactly the amount quantum mechanics predicts.

The gloves were not in the boxes.

The Idea That Is Actually Worth Having

Here it is, and it is worth the price of the whole chapter.

In an entangled system, **the joint state is completely specified while the individual states are not.**

Read that again, because it is not a poetic formulation. It is the mathematical situation. You can have maximal, complete, exhaustive knowledge of a two-particle system, leaving nothing whatsoever unknown about the pair, while each individual particle, considered alone, is in a state of maximal indefiniteness. There is no fact about particle A by itself. The facts live in the relationship.

Erwin Schrödinger, who coined the term entanglement in 1935, called this the characteristic trait of quantum mechanics, the one thing that most sharply separates it from classical thinking. Classical systems are made of parts with properties, and the whole is the sum of the parts. Entangled systems are not decomposable. The whole is real and the parts, in the strictest sense, are not fully there.

Now think about someone you have loved for a long time.

The Undecomposable Couple

Ask Nadia how she is doing and watch what happens to the question.

She cannot answer it. Not because she is evasive, but because after eleven years a large portion of her internal state is not indexed to her alone. Her mood on a given evening is a function of a variable that lives partly in another person. Her sense of how her career is going is inseparable from a set of decisions the two of them made jointly and revise jointly. Her account of who she is includes a chapter that cannot be written without him.

This is not codependence. This is the ordinary structure of a long attachment, and there is a substantial research literature on it.

Physiological linkage. In 1983 Robert Levenson and John Gottman brought couples into a lab, wired them for heart rate, skin conductance, and movement, and had them discuss a conflict. They found that the degree to which the partners' physiologies tracked each other during conflict predicted marital satisfaction, and the direction was the opposite of what you would guess: *higher* linkage predicted *lower* satisfaction and greater decline over the following years. Being locked into each other's arousal during a fight is not intimacy. It is a runaway feedback loop.

But linkage during calm, affiliative interaction is a different animal. Ruth Feldman's work on what she calls biobehavioral synchrony documents coordination of heart rhythms, oxytocin levels, gaze, and vocal rhythm between attached pairs, beginning with mother and infant and extending into adult romantic bonds. The coordination is real, measurable, and appears to be part of the mechanism by which attachment does its work.

Coregulation. David Sbarra and Cindy Hazan proposed in 2008 that a central function of a close relationship is regulatory: the presence of an attachment figure stabilizes your physiology, and over time your baseline recalibrates around their availability. This has a bleak corollary that explains something otherwise mysterious about breakups and bereavement. When the attachment figure disappears, you do not merely lose companionship. You lose a component of your regulatory system, and your body has to rebuild its stabilization from scratch, which is metabolically and psychologically expensive and takes far longer than anyone expects.

James Coan's neuroimaging work makes the same point from another angle. In a 2006 study, women anticipating an electric shock while inside a scanner showed markedly reduced threat response when holding their husband's hand, with the reduction proportional to marital quality. Coan's interpretation, which he calls social baseline theory, is that the human brain treats a reliable partner as part of its own resource base and adjusts its expected costs accordingly. You are not calculating threat as an individual. You are calculating it as a unit.

That is the closest legitimate analogue I know to the entanglement idea. Not a channel between you. A joint state, in which the individual states are less well defined than you assumed.

The Dark Version

Every good idea in relationships has a pathological form, and this one's is particularly nasty.

If the joint state can be well defined while the individual states are not, then it is possible to build a relationship in which the individuals have almost no definite state at all. Family systems theorists have a word for this: fusion, or enmeshment. It looks like closeness from the outside. It feels like closeness from the inside, at least at first. And it is characterized by a specific symptom, which is that neither person can locate a preference, an opinion, or an emotional reaction that is reliably their own.

Murray Bowen built much of his theory around the concept of differentiation: the capacity to maintain a self while in close emotional contact with others. Low differentiation does not mean distance. It means the opposite, an inability to be close without dissolving, which produces oscillation between fusion and cutoff because those are the only two available settings.

The physics gives us a precise way to say what has gone wrong. In a healthy long attachment, some variables are entangled and some remain locally defined. In fusion, everything is entangled, and when everything is entangled there is nobody to be in the relationship.

The practical diagnostic: can you locate an opinion, held sincerely, that your partner does not share, and can you say it out loud in their presence without either of you treating it as a threat to the union? If yes, you have local variables. If the honest answer is that you have not had a private opinion in some years, that is the finding, and it is not a small one.

Monogamy Of Entanglement

There is a theorem in quantum information theory with a name so on the nose that I hesitate to use it. It is called the monogamy of entanglement.

The result, formalized by Coffman, Kundu, and Wootters in 2000, says that entanglement cannot be freely shared. If two particles are maximally entangled with each other, neither can be entangled with a third at all. Partial entanglement can be distributed among several parties, but there is a strict conservation: what you share with one, you cannot also share with another. The total is fixed.

I want to be extremely careful with this one, because it is exactly the kind of result that gets weaponized.

It is not an argument about human relationships. Physics has no opinion about your domestic arrangements, and anyone who tells you a theorem about qubits settles a moral question is doing something dishonest. People build durable, ethical, loving lives in many configurations, and the research on consensually non-monogamous relationships does not show them to be systematically worse off.

What the theorem does offer, if you want it, is a clean way to think about a genuine constraint that is not about morality at all. It is about finiteness. Attention is finite. The hours available for the redundant world-building talk of the last chapter are finite. The capacity to be a stable regulatory presence for someone is finite. Whatever your arrangement, you are distributing a bounded resource, and the failure mode is always the same: promising an unbounded amount of it.

That constraint applies just as forcefully to a monogamous couple who have given the majority of their available presence to a job, a parent, and a hobby. The rival is rarely another person.

Practice: Locating The Joint Variables

This exercise takes about forty minutes and is best done separately and then compared.

Make three lists.

Mine. Things about you that would be substantially the same if this relationship ended tomorrow. Your taste in music. Your relationship with your brother. Your sense of humor. Your ambition.

Ours. Things that only exist in the pair and would not survive the separation intact. Your shared account of the year everything went wrong. The vocabulary. The way you divide a Sunday. The version of yourself that only appears when this specific person is in the room.

Contested. Things you are not sure about, or that you suspect you and your partner would categorize differently.

Then compare, and pay attention to two specific patterns.

If one person's "mine" list is dramatically shorter than the other's, you are looking at an asymmetry in differentiation, and it is usually invisible to both parties until it is on paper. The person with the short list has typically been praised for it, because a person whose individual state is undefined is extremely easy to live with, right up until they are not.

If the "ours" list is very short, that is the decoherence of the previous chapter showing up in inventory form. Two people with almost no joint variables are roommates with an unusual history.

What usually goes wrong: couples turn this into an argument about whether an item is correctly categorized. It is not a test. Disagreement about placement is the most useful output, because it means one of you has been treating something as shared that the other has been treating as their own, and that specific mismatch is where an enormous fraction of ordinary resentment lives.

What Actually Binds You

The correlation you feel when you both reach for the same joke at the same instant is not a signal passing between you. It is the product of a shared model, built out of nine thousand ordinary evenings, which I would argue is a more impressive achievement than telepathy and is certainly harder to come by.

But the structural insight survives all of the debunking, and it is this. Certain systems cannot be understood by examining their components, because the components do not carry the full story. The relationship is not something happening between two complete people. It is a state in which some of the facts about each person have relocated into the pair.

That is why endings are not subtraction. You do not go back to being who you were, minus a person. You go back to being someone whose individual state now has to be reconstructed for variables that were, for years, jointly held.

June knows this. She would put it differently. She would say that for eleven months after the funeral she could not have told you her own opinion of a film without checking it against a person who was not there.

Chapter Five: Nonlocality, Or Love Across A Distance

Dev is in Chicago. Marcus is in Lisbon. There are six time zones between them and have been for two years and four months, and the thing that nobody warned them about is that the hard part is not missing each other.

The hard part is that Dev learns about Marcus's day in a summary.

Marcus had a bad afternoon on a Wednesday in March. By the time they spoke, eleven hours later, he had processed it, filed it, and reduced it to a sentence. Dev received the sentence. Dev did not receive the eleven hours: the tightness in the shoulders, the second coffee, the particular quality of silence in the apartment, the way Marcus stood at the window without deciding to.

None of that is missing because of distance. It is missing because of bandwidth. And the difference between those two words is the entire subject of this chapter.

The Question That Bell Settled

Chapter Four ended with the experimental result, but I moved past its meaning quickly, and the meaning deserves its own treatment because it is one of the genuinely astonishing facts about the universe you live in.

Here is what the Bell tests establish. There exist correlations in nature that cannot be explained by any account in which distant things have independent definite properties and influences propagate no faster than light.

Read that carefully. It is not the claim that we have failed to find such an explanation. It is the claim that such an explanation is mathematically impossible, given the observed data. Bell's inequality is a ceiling that any local realistic theory must obey, and nature exceeds it.

So one of two deeply held assumptions has to go. Either distant systems do not have independent definite properties before measurement, or something nonlocal is going on. Physicists disagree about which to abandon, and the disagreement is respectable and ongoing, and it is a large part of why interpretation of quantum mechanics remains a live field a century in.

What nobody disputes is the experimental fact: the world contains correlations that distance does not degrade.

Two entangled photons show the same correlation whether they are separated by a meter or by the hundred and forty-three kilometers between La Palma and Tenerife, which Zeilinger's group demonstrated in 2012, or by the twelve hundred kilometers between a Chinese satellite and two ground stations, which Pan Jianwei's group demonstrated in 2017. The correlation does not weaken with separation. There is no falloff. There is no inverse square law for entanglement.

Every other connection you know about in physics gets weaker with distance. Gravity, electromagnetism, sound, light, all of them fade. This one does not.

What The Data Actually Say About Distance

Here is the part that surprises people.

Long-distance relationships are not, on average, worse. Laura Stafford's review of the literature, published as a book in 2005 and updated in subsequent work, found that long-distance romantic partners generally report levels of satisfaction, intimacy, trust, and commitment equal to or exceeding those of geographically close couples. A 2013 study by Crystal Jiang and Jeffrey Hancock in the

Journal of Communication found that long-distance partners disclosed more, and idealized their partners more, and reported more intimacy in their daily interactions than proximate couples did.

The stability findings are more complicated and honest reporting requires stating them. Breakup rates for long-distance couples are roughly comparable to those for close couples across most studies, but with a distinctive pattern: a substantial share of dissolutions cluster in the months immediately *after* reunion. Stafford and colleagues found in a 2006 study that about a third of couples who reunited broke up within three months of closing the distance.

That finding is the interesting one, and it explains itself with the two mechanisms we already have.

Idealization. Distance removes the ordinary friction data. You do not observe your partner being petty in traffic, or bad at conflict at eleven at night, or a person who leaves cabinet doors open. What you receive is curated by summary, and every summary is a small act of editing performed by someone who wants to be loved. When the distance closes, the friction data arrives in bulk, and it is experienced not as new information but as a discovery of concealment, which it usually is not.

Autonomy. Long distance provides an unusual amount of individual space alongside intense, high-attention connection. Some people discover on reunion that they did not want the arrangement they thought they wanted; they wanted this one, and they had it, and now it is gone.

So the honest summary is not that distance does not matter. It is that distance does not attenuate the bond, and it does change what the bond is made of, and the reckoning arrives later than you expect.

Bandwidth Versus Bond

Here is the distinction I want you to keep.

The bond is the joint state from the last chapter. The shared history, the private world, the fact that some of your variables live in the other person. Distance does not touch it. There is no evidence that attachment strength decays with kilometers, and the phenomenology of grief and separation across cultures suggests strongly that it does not.

The bandwidth is the rate at which the two of you exchange the material from which that joint state is built and maintained. Distance devastates bandwidth, and not in the obvious way.

The obvious losses are physical: touch, which does real regulatory work, as Coan's hand-holding study showed; shared physical presence, which delivers hundreds of small unremarkable data points an hour; sex, and the whole category of nonverbal repair.

But the loss that actually kills long-distance relationships is subtler, and Dev named it exactly. It is the loss of **unprocessed material.**

When you live with someone, you receive their raw state. You see the bad afternoon while it is happening, before it has been narrativized. You get access to the version of them that has not yet been made presentable, which is the version that a private world is built out of.

At distance, almost everything arrives processed. The call happens at a scheduled time. Both parties have had hours to formulate. Both are aware that the available window is limited, which creates pressure to use it well, which means important topics get prioritized and trivia gets cut. And trivia, as Berger and Kellner told us, is the engine.

This is why long-distance couples often report the strange experience of having deep, meaningful, emotionally serious conversations twice a week and still feeling that they are becoming strangers. The depth is not the problem. The depth is a symptom. All the shallow bandwidth has been eliminated, so every channel is doing heavy work, and nobody is doing the light work that actually constructs a shared reality.

What Works

Dev and Marcus figured out four things over two years, and I have seen versions of all four in the research literature and in practice.

Restore the trivial channel. This is the highest-value intervention and the one people resist because it feels unserious. Send the photograph of the strange thing on the sidewalk. Report the annoying interaction with the barista. Narrate the boring parts. The goal is not communication; the goal is volume of low-stakes shared reference.

Jiang and Hancock's finding that long-distance couples disclose more is real, but they disclose more *depth*, not more *width*, and width is what is missing.

Do things in parallel rather than only talking. Watch the same film at the same time with an open call. Cook the same recipe. Work quietly on separate tasks with video open and nobody speaking. Parallel activity generates shared reference and, critically, generates unprocessed material: Marcus can see Dev get irritated with a stubborn jar, which is worth more than a paragraph about how his week went.

The absence of pressure to perform is the active ingredient. Scheduled calls create a stage. Parallel presence creates a room.

Protect asynchronous intimacy. Voice memos rather than texts, because voice carries state that text strips. Letters, if you have the patience, because writing at length produces a version of a person that phone calls do not. Dev and Marcus write on Sundays, badly, at length, and both of them describe those letters as the most durable thing they have.

Set an end condition. This one is not optional and the research is consistent about it. Couples who have a concrete plan for closing the distance, with a date, do markedly better than couples in indefinite arrangements. Uncertainty about duration is corrosive in a way that duration itself is not. Two years with an end date is a project. Two years without one is a slow negotiation about whether this is your life.

And plan the reunion as an event requiring work, not a reward. Given that a third of dissolutions occur in the first three months after closing the distance, the reunion is the highest-risk phase of the entire relationship, and almost everyone treats it as the finish line.

The Other Kinds Of Distance

I want to widen this, because most of the readers of this chapter live with their partners.

Every mechanism above applies to couples separated by nothing but circumstance.

The partner who travels four days a week. The partner working nights while you work days. The couple in the same house who have not exchanged an unprocessed sentence in a year, because every interaction is scheduled, purposeful, and conducted through the summary layer.

Nadia and Wes live in a house of eleven hundred square feet and had, at the low point, less unprocessed contact than Dev and Marcus had across an ocean. Wes's bad afternoon in March also arrived as a sentence. It arrived as fewer than a sentence. It arrived as a man looking at his phone.

Proximity is not bandwidth. This is the sentence I would put on the cover if I could. Being in the room is a very good opportunity for bandwidth and is routinely wasted, and the fact that you are physically present makes the waste invisible, because you both assume that co-location is doing the work.

It is not. The work is done by unstructured, low-stakes, unedited contact, and you can go years without any of it while sharing a bed.

Practice: The Raw Channel

Three weeks. Both partners. Whether you live together or on separate continents.

Rule one: send unprocessed material at least twice a day. The standard is that it should feel slightly not worth sending. If you find yourself composing, stop and send the earlier draft. A photograph of nothing. A three-word observation. The thought you had that went nowhere.

Rule two: no response is required. This is the part that makes it work and the part everyone gets wrong. If every transmission demands a reply, you have created an obligation, and obligations produce resentment and then avoidance. The raw channel is broadcast, not correspondence. Marcus went four days without acknowledging anything Dev sent, and Dev kept sending, and it was fine, because the point was never the reply.

Rule three: one parallel session a week. An hour of doing something adjacent to each other with no agenda. In the same room or on an open call. Talking optional.

What usually goes wrong: week one is charming, week two feels forced, and week three is where most people quit. Push through week two. The forced feeling comes from the fact that you are consciously performing a behavior that used to be automatic, and consciousness of it fades once the volume gets high enough that you stop curating.

The second failure: one partner sends and the other does not, and the sender begins to keep score. If you find yourself counting, say so out loud, in plain words, quickly. That is the observer effect from Chapter One arriving on schedule, and it is far easier to interrupt at item four than at item four hundred.

What The Photons Would Say

I have been careful throughout this chapter, so let me spend the last paragraph less carefully.

The Bell experiments tell us that we live in a universe where correlation is not built out of proximity. Whatever ties two entangled particles together, it is not a thread, not a signal, not a force running between them through the intervening space, and it does not care how much space there is.

I do not think that fact is about your relationship. I think it is a fact about photons.

But I notice that almost every intuition we have about human closeness is spatial. We say we feel distant. We say we have drifted apart. We say we want to be close. The metaphors are all geometry, and the geometry is misleading, because the thing that actually determines whether two people remain in a correlated state has almost nothing to do with the distance between their bodies and almost everything to do with the volume and rawness of what passes between them.

Dev and Marcus are six time zones apart and are doing better than a great many couples who share a bathroom.

That is not spooky action at a distance. It is two people who understood, earlier than most, that proximity was never the mechanism.

Chapter Six: Coherence, Or Two Systems In Phase

There is a lightbulb in your kitchen and a laser pointer in your desk drawer, and the difference between them is not power.

A hundred-watt incandescent bulb radiates far more energy than a laser pointer, which typically runs at about a thousandth of a watt. The bulb will warm a room. The pointer will not warm anything. And yet the pointer will put a visible dot on a wall three hundred meters away, and will cut through fog, and a slightly larger version of the same device will cut through steel, while the bulb's light is exhausted in a few meters of ordinary air.

The bulb has more energy. The laser has coherence.

I have come to think this is the most underrated idea available to couples, and the one most often confused with something else, which is agreement.

What Coherence Means

Light is a wave. Every wave has a phase, which is simply where it is in its cycle at a given instant: at a crest, at a trough, or somewhere in between.

In a lightbulb, each excited atom drops to a lower energy state and emits a photon whenever it happens to, independently of every other atom. The emissions have random phases, random directions, and a broad spread of frequencies. The waves are constantly interfering with each other, sometimes adding, sometimes cancelling, in a chaotic pattern that averages out. The result is a lot of energy going everywhere in an uncorrelated mess.

In a laser, the emissions are stimulated rather than spontaneous. A passing photon triggers an excited atom to emit a second photon with the *same* frequency, the *same* direction, and the *same* phase. That photon triggers another. The result is an enormous population of waves marching in step.

And here is the fact that makes this worth a chapter. When N wave sources are incoherent, the intensities add: total brightness is proportional to N. When N sources are coherent, the *amplitudes* add before squaring, and the intensity is proportional to N squared.

Ten thousand incoherent emitters give you ten thousand units of brightness in a given direction. Ten thousand coherent emitters give you a hundred million.

The energy is the same. The organization is different. Coherence buys you a superlinear return, and it does not cost anything except the coordination itself.

Coherence Is Not Sameness

This is the distinction everything else in the chapter depends on, so I want to state it precisely.

Two waves are coherent when they maintain a **constant phase relationship** over time. That does not mean they must be in phase. Two waves that are permanently a quarter cycle apart are perfectly coherent, and their interference produces a stable, predictable pattern. What destroys coherence is not difference. It is *drift*: a phase relationship that wanders unpredictably.

You can see this in the physics of interference. The stripes in the double slit experiment require the two paths to be coherent with each other. They do not require the two paths to be identical; they are not identical, they have different lengths, and the difference in length is exactly what produces the pattern. The pattern exists *because* of the difference, and it is stable because the difference is constant.

An incoherent source produces no stripes, not because its waves are too different from each other but because the relationship between them will not hold still.

Now transfer this to two people, and notice what it rules out.

The popular theory of relational alignment is compatibility: find someone who wants the same things you want. The physics suggests a different criterion. What you need is not that your partner's phase equals yours. What you need is that the relationship between your phases is stable and known.

Nadia wants to move to Portland. Wes does not. That is a phase difference, and it is permanent, and it has been permanent for six years. It is not what is hurting them. What was hurting them was that neither of them knew where the other actually stood, because the subject had become dangerous and was therefore approached only obliquely, and each of them was working from a model of the other that was five years old and never verified.

They are not going to agree about Portland. They can be coherent about Portland. Those are different projects, and only one of them is achievable.

The Superlinear Couple

Let me make the first fact concrete, because it is not merely poetic.

Two people who have negotiated a stable division of domestic responsibility do not save the time of one person. They save considerably more, because each of them stops paying the cognitive cost of tracking, deciding, and renegotiating. That overhead is invisible and enormous. Research on what sociologists call the mental load, and Allison Daminger's 2019 work on cognitive labor in particular, documents that the anticipating, monitoring, and deciding components of household work are both substantial and unevenly distributed, and that they are not captured at all by time-use surveys that measure execution.

Coherent couples eliminate the overhead by making the relationship stable rather than by making the work equal. This is why a rigid and slightly unfair division often outperforms a flexible and theoretically fair one, and why "we just figure it out as we go" produces more resentment than a system nobody loves.

The same applies to money, to holidays, to parenting philosophy, to how you handle each other's difficult relatives. The value is in the constancy of the phase relationship, not in the values themselves.

And there is a second-order effect. When the relationship is stable, each person can *predict* the other, and prediction is what allows you to act on the other's behalf without checking. Wes buys the coffee Nadia likes without a text. Marcus knows which of Dev's complaints want sympathy and which want to be forgotten by morning. Every one of those is a small superlinear return: a coordination achieved with zero communication cost, purchased by years of stable phase.

That accumulated predictive model is most of what people mean when they say a long relationship is easy. It is not easy. It is amortized.

Shared Reality

There is a research construct that maps onto this well enough that I want to name it.

Curtis Hardin and Tory Higgins introduced the idea of shared reality in the 1990s: the proposition that our sense that our own experience is valid depends substantially on having it confirmed by another person. What is merely a private impression becomes a fact about the world when someone else has it too. Subsequent work by Maya Rossignac-Milon and Higgins has extended this specifically to close relationships, developing a measure of what they call generalized shared reality and finding it associated with relationship quality, closeness, and the sense that a relationship is meaningful.

This is coherence at the level of interpretation. Two people who have a stable, mutually confirmed account of what things mean are not simply more comfortable. They are getting a superlinear epistemic return: each of them is more confident, more decisive, and less anxious than they would be alone, because their read on the world has been ratified.

The failure mode is the one you would predict from the physics and it is worth watching for. A couple with very high shared reality and very low external contact can ratify each other into positions that no third party would endorse. Coherence amplifies whatever it is coherent about. A laser will happily deliver a beautifully organized beam of complete nonsense.

The Michelangelo Effect

The most striking finding in this area comes from Caryl Rusbult and colleagues, who documented what they called the Michelangelo phenomenon in a series of studies from the late 1990s onward.

The premise: each of us has an ideal self, a version we are trying to become. Partners, through the way they perceive and behave toward us, either sculpt us toward that ideal or away from it. Rusbult's finding was that when a partner's perception and behavior are aligned with your ideal self, you actually move toward it over time, and both personal wellbeing and relationship quality improve.

The distinction that makes this useful is between affirmation and what Rusbult called the Pygmalion pattern of sculpting someone toward *your* ideal for them rather than theirs. The second one looks identical from the outside and produces the opposite result.

In the coherence frame, affirmation is phase alignment with someone's own trajectory. You are not trying to make them match you. You are trying to make the relationship between where they are going and how you see them stable and reinforcing, so that their movement and your perception add rather than cancel.

Tom does this well and does not know he does it. Priya has a half-serious ambition to teach, which she mentions rarely and dismisses when she does. Tom has never argued with the dismissal. He has simply, for two years, treated it as an existing fact about her, referring to it in passing, asking practical questions as though the decision were already made. She applied in March.

Where Coherence Is The Wrong Goal

I should say plainly that alignment is not always the answer, because a book that recommends one thing for two hundred pages is not a book, it is a pamphlet.

Some phase differences should not be stabilized, they should be resolved. Whether to have children is not a difference you can coherently maintain. Some couples try, holding it as a permanent open question with a stable relationship to it, and it is one of the most reliably destructive arrangements I know of. There is a small set of questions where the answer must actually be settled, and treating them like Portland is a way of avoiding a decision for a decade.

Coherence can be purchased at too high a price. A couple can achieve a beautifully stable phase relationship in which one person's phase is simply set to match the other's. That is not coherence between two systems. That is one system and an accessory. The check from Chapter Four applies: if one of you no longer has locally defined variables, the stability you have achieved is not the good kind.

Novelty requires incoherence. Arthur Aron's self-expansion model, supported by a well-replicated body of work, finds that couples who engage in novel and arousing shared activities show increases in relationship quality, while couples who do only pleasant familiar things do not. Novelty is by definition a disruption of the established pattern. A relationship that has optimized entirely for stable phase has optimized away the mechanism by which it grows.

The resolution is that you want coherence in your infrastructure and incoherence in your experience. Stable agreements about money and labor and what things mean. Unstable, surprising, occasionally disorganizing things to do together.

Most couples get this exactly backwards. They have a chaotic and constantly renegotiated infrastructure and a completely predictable set of activities.

Practice: The Phase Audit

Once a year, ninety minutes, and it should be uncomfortable.

Make a list of the ten or twelve subjects that most affect your shared life. Money. Work hours. Where you live. Sex. Extended family. Children. Health. Faith or its absence. Retirement. What a good weekend looks like.

For each, both of you write down two things independently: **where you actually stand**, and **where you believe your partner stands.**

Then compare.

The gaps between column two and your partner's column one are the finding. That is where you are running on a stale model. Most couples discover between two and four of these, and at least one of them will be several years out of date and will have been quietly generating friction the whole time.

Then, for each subject, choose a category out loud:

Aligned. Same direction. Nothing to do but confirm it.

Stably different. A real difference that does not need resolving. Portland goes here. The work is to state it explicitly, so that neither of you has to keep re-discovering it, and so that neither of you interprets the other's occasional wistfulness as a new campaign.

Requires resolution. A difference that cannot be held indefinitely without damage. These get a date, and the date goes on a calendar, and you do not leave the conversation without it.

What usually goes wrong: everything gets sorted into "stably different," because that category is comfortable and requires nothing. If you have more than two items in it, you are probably using it as a hiding place. The test is whether you can imagine holding the difference for ten more years without either of you becoming someone smaller. If not, it belongs in the third category.

The second failure is doing this during a fight. It does not work during a fight. Schedule it when nothing is wrong, which is precisely when nobody wants to, which is why almost nobody does it.

The Laser And The Bulb

I want to end with the comparison I opened with, because I think it contains the actual argument.

The lightbulb is not doing anything wrong. It is radiating enthusiastically in all directions with a great deal of energy. It is generous, chaotic, warm, and completely unable to reach across a room.

Plenty of couples are lightbulbs. There is real love in the house, considerable effort, genuine feeling, and it is being spent in every direction at once with no stable relationship between the parts, and the effect at any distance is nothing.

The laser is not trying harder. It is not more in love. It has a thousandth of the energy.

It just knows what its parts are doing relative to each other, and it keeps knowing, and that is enough to cut steel.

PART THREE

Indeterminacy: What Cannot Be Pinned Down

Chapter Seven: Superposition, Or Holding Two True Things At Once

June says the sentence in a coffee shop, quietly, and then looks alarmed at herself.

"I miss him every day and I am relieved he is gone."

Then, immediately: "That is a terrible thing to say."

It is not a terrible thing to say. It is an accurate report of a state that our vocabulary handles so badly that the person experiencing it assumes she must be defective. She nursed him for nineteen months. She loved him for thirty-four years. Both halves of the sentence are true, at full strength, at the same time, and she has spent two years trying to determine which one is the real feeling underneath the other one.

There is no underneath. That is the entire content of this chapter.

The Thing Superposition Is Not

Let me clear away the cartoon first.

You have been told that superposition means a particle is in two places at once. This is the standard formulation and it is not quite right, and the way in which it is not right turns out to be the useful part.

A quantum state is described by something called a wave function. Superposition means that a state can be written as a combination of other states. An electron's spin, measured along the vertical axis, comes out either up or down, never anything else. Before you measure, the electron can be in a state that is a combination of up and down.

The cartoon says: it is secretly both up and down, and measurement reveals which. That is wrong in a way that matters. The state is not a mixture of two definite conditions that we are ignorant about. It is its own thing, a single definite state, which happens to have no definite value for the vertical spin question.

And here is the fact that makes superposition genuinely interesting rather than merely weird.

A superposition in one basis is a definite state in another.

An electron in an equal superposition of up and down along the vertical axis is in a perfectly definite, unambiguous state along the horizontal axis. It is not confused. It is not half of anything. It has a completely sharp value for a different question. What you were calling indeterminacy was an artifact of the question you were asking.

Change the axis and the ambiguity vanishes.

I have never encountered a more useful idea for human emotional life, and I do not think it is a coincidence that the same insight appears in contemplative traditions that had no access to physics.

The Fine Print

The mathematics is prettier than you would expect and I will do it in words, once.

A quantum state lives in a space with several directions available, and choosing a measurement is like choosing which axes to describe a vector against. A vector pointing diagonally is "a combination of north and east" relative to those axes. Rotate the axes and the same vector points straight along one of them. Nothing about the vector changed. The description changed.

That undersells the weirdness in one respect, which is that quantum amplitudes are complex numbers and can cancel each other outright. Superposition is not bookkeeping; it leaves fingerprints on a wall.

But the part I want you to steal is the rotation. **Ambiguity is relative to the question.** Ask a different one and the same state can be perfectly sharp.

People have been arriving at that from meditation cushions for three thousand years. It is nice to have it confirmed by an electron.

Rotating The Axis

June's problem is that she has been asking a vertical question.

Do I miss him, or am I relieved? Along that axis, her state has no definite value, and every attempt to resolve it produces the same non-answer, which she experiences as evidence of something wrong with her.

Now rotate.

Ask instead: *What was the last nineteen months?* And the answer is sharp, immediate, and entirely unambiguous. It was the hardest thing she has ever done. It was a period in which she loved someone and cleaned him and lost sleep and lost herself and was frightened and was necessary, continuously, for five hundred and eighty days.

Along that axis there is no contradiction at all. Missing him and being relieved are not two competing readings of her state. They are two projections of one thing onto an axis that cannot represent it.

This is not a rhetorical trick, and I want to defend it, because it sounds like one.

There is a well-supported finding in affective science on what Lisa Feldman Barrett calls emotional granularity: the degree to which a person parses their emotional experience into fine-grained, differentiated categories. People with high granularity, who distinguish irritated from resentful from disappointed rather than filing all three as "bad," regulate more effectively, drink less in response to stress, and are less likely to respond aggressively. Matthew Lieberman's work on affect labeling found that putting feelings into words is associated with reduced amygdala response and increased prefrontal activity, which is a plausible mechanism.

What both lines of work suggest is that the *description* is not passive. Finding the right axis is doing something.

June's "I miss him and I am relieved" is a low-granularity description forced through a binary question. "I spent nineteen months doing something enormous and it ended" is a higher-granularity description along a better-chosen axis, and it does not feel like a contradiction, and it is the same state.

Ambivalence Is Not A Defect

Our culture treats mixed feelings as a temporary condition to be resolved. You are supposed to figure out how you really feel. The mixed state is understood as a waiting room.

Two lines of evidence say otherwise.

Emotional complexity is normal and increases with maturity. Research on emotional complexity, including work by Laura Carstensen and colleagues on age-related changes, finds that older adults report co-occurrence of positive and negative emotion more frequently than younger adults, and that this poignancy is associated with better emotional regulation rather than worse. The capacity to hold mixed states appears to be something people develop, not something they grow out of.

Cultural framing changes the experience. There is a substantial cross-cultural literature, associated with Yuri Miyamoto, Shinobu Kitayama, and others, finding that people in East Asian cultural contexts, which tend toward dialectical rather than linear reasoning about contradiction, report mixed emotions more readily and experience less distress about them than people in Western contexts. Same states. Different tolerance.

The relevant caveat, and I want to give it fairly: not all ambivalence is benign. Bert Uchino's research on ambivalent social ties, relationships that are both supportive and aversive, finds that they are associated with worse cardiovascular outcomes than either consistently positive *or* consistently negative relationships. A friend who is sometimes wonderful and sometimes cutting is harder on your body than a reliably unpleasant one, presumably because unpredictability prevents you from settling into any stance.

So there are two different things, and it is important not to confuse them.

Ambivalence about a stable situation is usually fine, often accurate, and typically a sign that you are perceiving something complicated correctly. June's state is this kind.

Ambivalence generated by an unstable situation is a signal, not a state to be accepted. If you cannot tell whether you feel safe with someone because their behavior oscillates, the ambivalence is a correct readout of an actual oscillation. Rotating the axis will not help. The axis is fine. The situation is the problem.

Telling these apart is the skill. The rough test: has the underlying situation been stable for a long time while your feelings about it remain mixed? That is superposition. Do your feelings track a partner's shifting behavior with a lag of about two days? That is a measurement of something real, and you should take the reading seriously.

The Argument You Cannot Win

The single most common place superposition shows up in couples is the demand for a single position.

Priya, at the end of a difficult month: "I want to be with you and I feel trapped."

Tom hears a contradiction and does what any reasonable person does with a contradiction, which is to try to resolve it. Which do you mean. Which is more true. If you feel trapped, do you want to leave.

And now Priya has to pick one, and whichever she picks will be a distortion, and the conversation will proceed from a false premise into forty minutes of increasingly unrecognizable territory. By the end she will have defended a position she does not hold, because she was forced to project a two-component state onto a one-dimensional question and then argue about the projection.

The move that prevents this costs one sentence, and it is the hardest sentence in the book to actually say in the moment.

"Both of those can be true. Tell me more about the second one."

Not: which is it. Not: but you just said. Not the perfectly reasonable, entirely fair, relationship-destroying observation that these two statements are logically in tension.

Because they are not, actually, in tension. They are in tension along the axis Tom is using, which is *do you want this relationship, yes or no*. Along the axis Priya is actually reporting from, which is something like *what has this month cost me*, they are one coherent statement.

Practice: The Rotation

This is for use when you are stuck in an internal contradiction that has not resolved for a while.

Step one. Write the contradiction in its binary form, exactly as it appears in your head. "I want the job and I want to stay here." "I love her and I am bored." "I miss him and I am relieved."

Step two. Notice that the binary form is a question you are asking. Write the question explicitly. Usually it is some version of "which do I really want" or "which do I really feel." The word *really* is the tell. It appears whenever someone is looking for a hidden variable.

Step three. Generate three different questions that the same emotional material could be an answer to. Not rephrasings. Genuinely different axes. Some that work:

What have I been doing for the past year, and what has it cost?

What am I afraid will happen if I say this out loud?

What did I want when I started, and is that person still here?

What would I tell someone I love who described this to me?

Step four. Answer each one in a paragraph. Look for the one where you find yourself writing quickly and without qualification.

That is the axis along which your state is definite. You will usually know it within two sentences.

What usually goes wrong: people treat this as a technique for figuring out which side of the binary is correct, and then feel that it failed. It is not a decision procedure. It will not tell you whether to leave your marriage. What it does is get you out of a question that has no answer and into one that does, which is a precondition for any decision but is not itself a decision.

The other failure: doing this exercise once, getting a clean rotation, and expecting it to hold. States evolve. You may have to rotate again next month. That is not a sign the first rotation was wrong.

For The Partner Of Someone In Superposition

A short section, because this is the harder role and it gets less attention.

If someone you love says two contradictory things about you, the demand to know which one is real is completely understandable and almost always destructive.

What you actually need to know, and what you are entitled to ask, is not which feeling is real. It is what they intend to do. Those are separable, and conflating them is where the damage happens.

"I love you and I am not sure I want this life" is a state. It may persist for a long time. It is not answerable.

"Are you staying through the end of the year, and are you going to work on this with me" is a question about action, and it has an answer, and you are owed it.

The generosity is in accepting the ambiguity of the internal state. The self-respect is in requiring clarity about conduct. People who give the first without the second end up waiting for years. People who demand the first as the price of the second end up with a partner who has learned to report a single clean feeling that is not true.

June's husband, near the end, said something she has never told anyone. He said he was ready to go and he was furious about it and he did not want either of those written on anything.

She did not ask him which one he meant. It is, she says, the thing she is proudest of.

Chapter Eight: The Uncertainty Principle, Or Closeness And Freedom As A Conjugate Pair

Every couple has one argument. Not one argument at a time; one argument, which wears different costumes for decades.

For a large fraction of couples, the argument is this one. One person wants more contact and the other wants more room, and both of them experience the other's want as a criticism, and neither of them can concede because neither of them is wrong.

Nadia and Wes have been having it since their second year. It has appeared as an argument about how often his mother visits, about whether he texts back during the day, about the fishing trip, about whether "I need an hour" is a reasonable thing to say to a person who has been alone with two children since six a.m. Each specific instancc was resolved. The argument was never touched.

I want to convince you of something that will not fix this and may make it easier to live with: the tension is not a symptom of anything. It is structural, it is permanent, and the two things you are fighting over are related to each other the way position and momentum are.

What Heisenberg Actually Said

In 1927 Werner Heisenberg published the relation that carries his name, and he explained it with the microscope argument I mentioned in Chapter One: to locate an electron you must hit it with a photon, and the photon disturbs it.

That explanation is intuitive, memorable, and not the right account of what the principle is.

The modern understanding is cleaner and much stranger. Position and momentum are what physicists call conjugate variables, and the relationship between them is mathematical rather than mechanical. It comes from the fact that a quantum state is a wave, and that the momentum description of a wave is its Fourier transform.

Here is what that means without the mathematics. Think about a sound. A pure, sustained tone, one single frequency, has to last a long time; a frequency is a rate of oscillation and you cannot establish a rate from a very short sample. Now think about a click, a sound sharply located in time. A click is not a pure tone at all. Analyze it and you find it contains a broad smear of frequencies.

This is not a limitation of your ear or your equipment. It is a property of waves. **A wave that is sharply defined in time is necessarily spread out in frequency, and a wave that is sharply defined in frequency is necessarily spread out in time.** There is no clever engineering that beats this, because it is not about measurement. It is about what a wave is.

Position and momentum stand in exactly this relationship. A quantum state narrowly localized in space is, as a matter of mathematical necessity, a superposition of many momenta. Not because we disturbed it. Because a localized wave is built out of many wavelengths.

So the uncertainty principle is not a statement about our clumsiness. It is a statement that certain properties come in pairs, and that sharpness in one *is* spread in the other, and that this is a feature of the object rather than the observation.

The Fine Print

Two things, and the second one is the gift.

Heisenberg's original story was about clumsiness: to see an electron you have to hit it with a photon, and the photon shoves it. Memorable, intuitive, and not the real reason. The real reason is that a wave sharply defined in one quantity is *built out of* a spread in the other. It is not damage. It is architecture.

And here is the gift. **The principle sets a floor on the product, not on either factor.** You can have very sharp position and very fuzzy momentum, or the reverse, or anything in between. The total is fixed; the distribution is yours.

Physicists do this on purpose and call it squeezing. LIGO uses squeezed light to detect gravitational waves, deliberately dumping uncertainty into a quantity nobody cares about so it can read the one that matters.

You are going to do the same thing with your Saturdays.

The Oldest Dialectic

Relationship scholars got here independently and about sixty years after Heisenberg.

Leslie Baxter and Barbara Montgomery's relational dialectics theory, developed through the 1980s and 90s, holds that close relationships are organized around ongoing tensions rather than around achieved states. The central one they identify is autonomy and connection. Others include openness and closedness, novelty and predictability.

The critical claim in their framework, and the reason I find it more honest than most relationship theory, is that these tensions are not problems to be solved. There is no equilibrium point at which a couple has correctly balanced autonomy and connection and can stop. The tension is the ongoing condition of being close to someone, and relationships are conducted *within* it rather than being repaired *out* of it.

If you have spent years assuming your relationship has a bug because this keeps coming up, that assumption is the actual problem. It is not a bug. It is the specification.

The Dependency Paradox

There is a finding here that seems to contradict everything I have just said, and understanding why it does not is the most useful thing in this chapter.

Brooke Feeney published a study in 2007 following married couples over several years, examining whether relying on a partner undermines independence. The intuition says yes: the more you lean, the less you can stand.

The data said the opposite. Partners who were more responsive to a spouse's needs for support had spouses who subsequently showed *more* independent functioning: more goal pursuit, more autonomous exploration, more self-sufficiency. Feeney called it the dependency paradox, and it follows directly from Bowlby's attachment framework, in which a reliable secure base is precisely what makes exploration possible. An infant with a responsive caregiver ventures further from the caregiver, not less far.

So intimacy and autonomy are not straightforwardly opposed. Reliable closeness *produces* freedom.

How does that square with the conjugate pair?

Like this. The uncertainty relation is about two things being sharply defined *at the same moment*. The dependency paradox is about a longitudinal effect. Over months and years, secure connection increases autonomous capacity. In any given hour, you still cannot be both maximally merged with another person and maximally free of them.

Both are true, and they operate on different timescales, and couples confuse them constantly. Wes's request for an hour alone on a Saturday is a momentary distribution choice. Nadia's fear is about the longitudinal trend. They are answering different questions and each thinks the other is being obtuse.

Naming the timescale usually defuses about half of it. "I am asking for this afternoon, not for a different marriage" is a sentence that has ended more fights than it has any right to.

Demand And Withdraw

I have to name the pattern, because it is the mechanism by which this particular tension goes from uncomfortable to corrosive.

Andrew Christensen and Christopher Heavey documented the demand-withdraw pattern in a series of studies beginning in the 1980s. One partner pursues, criticizes, presses for discussion, requests change. The other avoids, defends, goes quiet, leaves the room. It is one of the most reliable predictors of relationship distress and dissolution in the entire literature, and it has been replicated across cultures.

The important structural fact is that it is self-amplifying. Pursuit produces withdrawal. Withdrawal produces more pursuit. Neither person is initiating; each is responding to the other, and each experiences their own behavior as reactive and the other's as the cause. Ask them who started it and both will answer honestly and differently.

There is a common gendered finding, wife-demand and husband-withdraw being more frequent than the reverse, but the better-supported explanation is not about gender essentials. Christensen and Heavey found that the pattern tracks who wants change: whichever partner is seeking a change in the status quo tends to occupy the demand role. When the topic was one where the husband wanted change, the roles reversed.

What breaks it is almost never an agreement about the substance. It is a structural intervention:

The withdrawer commits to a return time. Not "I need space," which reads as abandonment and triggers more pursuit. "I need forty minutes and then I will come find you," which is space plus a guarantee. The guarantee is the entire content. John Gottman's research on physiological flooding supports the timing: once heart rate exceeds roughly a hundred beats per minute, effective processing stops, and it takes something on the order of twenty minutes to come back down. A break is not avoidance. A break with no stated end is.

The pursuer stops at the first sign of flooding. This is much harder, because the pursuer's urgency is real and the fear of being left with the thing unresolved is real. But pressing a flooded person accomplishes nothing except teaching them that the only reliable way to end the pressure is to leave the room.

Nadia is the pursuer. Wes is the withdrawer. They have never once had a productive conversation past the ninety-minute mark, in eleven years, and they kept having them anyway, because stopping felt like losing.

Squeezing: Choosing Your Distribution

Here is the practical frame, and it is the reason I put the physics in.

You cannot have maximal closeness and maximal independence. That is the floor on the product. But you can decide, on purpose, together, where to spend the uncertainty.

Different couples squeeze differently and there is no correct answer.

Some couples run very high connection and very low autonomy: shared friends, shared work, little time apart, high mutual visibility. This works, for some people, extremely well. Its failure mode is the fusion of Chapter Four.

Some run high autonomy and lower moment-to-moment connection: separate social worlds, separate travel, long periods of independent absorption, reconnection in concentrated bursts. This also works. Its failure mode is decoherence, Chapter Three, arriving quietly.

The couples who suffer most, in my observation, are not the ones at either extreme. They are the ones who never chose, and are therefore running an implicit distribution that neither of them endorses, negotiated instance by instance, with every single request treated as a referendum.

That is the actual cost of not choosing: every Saturday becomes a fresh argument about the fundamental nature of the marriage.

Practice: The Distribution Conversation

This should happen once, take about two hours, and be revisited annually. It is not a fight and should not be scheduled anywhere near one.

Part one, separately. Each of you writes down, honestly, your ideal ratio for the coming year. Concretely, not abstractly. How many evenings a week alone. How many nights away is fine. How much of your social life should be shared. How much you want to know about the other's day, in minutes. How reachable you want to be during work.

Use numbers. Abstractions like "some space" and "enough closeness" are how couples avoid this conversation for decades.

Part two, together. Read them to each other and do not respond for five minutes. Not a discussion; a reading. The urge to defend or explain will be very strong. Sit in it.

Part three. Find the gaps and name them as a distribution problem rather than a values problem. "You want four evenings and I want one" is a distribution. "You do not care about this family" is a values accusation, and it is almost never accurate, and it makes the distribution unnegotiable.

Part four. Choose an explicit distribution for the next three months. Write it down. It will feel absurdly mechanical. It will also mean that when Wes goes fishing in April, he is executing an agreement rather than making a unilateral withdrawal, and Nadia is watching a plan happen rather than watching a man leave.

What usually goes wrong: the person who wants more autonomy under-reports it, because they have learned that asking for space causes pain, and so they ask for less than they need and then take the rest by drifting. This is the single most common failure of this exercise and it produces a settlement that quietly collapses in six weeks. If you are the low-contact partner, the useful discipline is to state the true number even though it will land badly, because a number that lands badly and gets negotiated is far better than a number that is agreed to and then silently exceeded.

The mirror failure: the person who wants more connection inflates the number to leave room for concession, and then feels cheated when they get what they actually wanted.

Both failures are the same failure. Someone is bargaining instead of reporting.

What You Get For Accepting It

I promised at the start that this would not fix anything, and I want to end honestly.

Nadia and Wes still have the argument. They had it in June. What has changed is the frame around it, and the frame changes the damage more than you would think.

They no longer treat it as evidence that they are mismatched, because they understand that the tension is a property of being close to another person rather than a property of these two particular people. They no longer expect a final resolution, so a recurrence is not a failure of everything they have built. And they have an explicit distribution, so the individual instances are administrative rather than existential.

That is a real gain. It is not the gain people want. People want the tradeoff to be an error they can correct, and it is not, and the sooner you stop trying to correct it the sooner you can start choosing it.

The floor on the product is fixed. Everything else is yours.

Chapter Nine: Wave And Particle, Or The Two Faces Of Love

Priya has a question she is embarrassed by, and she asks it in the tone people use for confessing something shameful.

"Is this still love, or is it just a habit that we are both being polite about?"

Two years ago she could not eat. She thought about Tom during meetings. There was a night in a stairwell that she still cannot describe without her voice changing. Now there is a Thursday routine and a shared calendar and a comfortable silence at breakfast, and she does not know whether the comfortable silence is the thing love turns into or the thing that replaces it after love leaves.

This is, I think, the most consequential question in the book, because getting it wrong ends relationships that did not need to end. The number of people who have left a good partnership because it stopped producing the stairwell is not small.

The Question That Broke Classical Physics

For two hundred years, light was a wave. Young's stripes had settled it, Maxwell's equations had explained it, and the matter was closed.

Then in 1905 Einstein explained the photoelectric effect, which is the phenomenon that light striking a metal knocks electrons loose. The puzzle was that whether electrons came out depended on the light's *frequency*, not its intensity. Dim blue light liberated electrons. Blindingly bright red light did not. On a wave picture this is incomprehensible; enough energy should eventually do the job regardless of color.

Einstein's resolution was that light arrives in discrete packets whose energy depends on frequency. A single red packet does not carry enough energy to free an electron, and you cannot combine two of them. Light is granular. It is a particle.

So light was a wave, demonstrated conclusively, and light was a particle, demonstrated conclusively. In 1924 Louis de Broglie proposed that matter has the same double character, and within a few years electrons had been shown to diffract, confirming it.

For about twenty years this was called wave-particle duality and it was regarded as a deep paradox at the heart of nature.

The Fine Print

Almost nobody tells you this part, and it changes everything.

Modern physicists mostly dislike the phrase "wave-particle duality," because a photon is not two things taking turns. It is one thing: an excitation of a field, described by a single mathematical object, which is neither a little ball nor a ripple on a pond. Both words are borrowed from our enormous slow world and forced onto something that was never either.

The photon is not switching. *We* are switching, between two descriptions that were always slightly wrong.

There is even a dial. In the 1990s Berthold-Georg Englert derived a precise relation trading interference sharpness against which-path information, continuously. You can have some of each.

So the lesson is not that love has two natures. It is that love has one nature and we own two words, and the contradiction was never in the thing. It was in the vocabulary.

Passionate And Companionate

The relationship literature has its own version of this dichotomy, and it has the same problem.

Elaine Hatfield and colleagues distinguished passionate love, characterized by intense longing, physiological arousal, preoccupation, and idealization, from companionate love, characterized by deep affection, trust, and intertwined lives. Ellen Berscheid and others developed the distinction further, and it became the standard framework.

The usual narrative that grew up around it goes: passionate love is a temporary state lasting somewhere between twelve and thirty-six months, driven by novelty and uncertainty, which then either transitions into companionate love or does not, and the transition is a decline from something thrilling to something warm.

Two pieces of evidence complicate that story considerably.

Romantic love does persist. Bianca Acevedo and Arthur Aron published a meta-analysis in 2009 examining long-term relationships and found that romantic love, defined as intensity, engagement, and sexual interest, was robustly associated with satisfaction in long-term relationships. What declined was not romantic love but *obsession*: the preoccupation, anxiety, and intrusive thinking characteristic of early-stage attachment. Their conclusion was that romantic love minus obsession is entirely compatible with a twenty-year marriage, and that people who assume the whole package must fade are conflating two separable components.

A companion neuroimaging study by Acevedo, Aron, Fisher, and Brown in 2012 scanned people who reported still being intensely in love after an average of twenty-one years of marriage, and found activation in dopamine-rich reward regions similar to that seen in early-stage romantic love, with additional activity in regions associated with attachment and pair bonding, and without the anxiety-associated activation present in the early-stage groups.

The categories are not sequential stages. The framing of passionate love as a phase you exit implies a one-way door. What the evidence actually supports is that both are available at any point, with different frequencies and different triggers, and that the conditions producing each are largely controllable.

Which brings us back to the physics.

Which Experiment Are You Running

Passionate and companionate are not two substances, one of which you have and the other of which you had. They are two descriptions of one bond, and which one shows up depends on the conditions.

The conditions that produce the passionate description are reasonably well characterized: novelty, uncertainty, physiological arousal, obstacle, and separation. Aron's self-expansion work is the strongest evidence here. In a series of studies, couples randomly assigned to novel and arousing joint activities showed increases in relationship quality relative to couples assigned to pleasant but familiar activities. Arousal appears to be part of the mechanism, which also explains the famous 1974 Dutton and Aron bridge study, in which men who had just crossed a frightening suspension bridge showed more romantic interest in an interviewer than men crossing a stable one.

The conditions that produce the companionate description are the opposite: predictability, safety, routine, ease, and continuous availability.

Notice that these two condition sets are not compatible in the same moment. This is complementarity again, in Chapter Two's sense: two arrangements, each producing a valid and complete description, mutually exclusive in application.

So the honest answer to Priya's question is that she has not lost anything. She has been running the companionate arrangement continuously for eighteen months, because that is what a shared calendar, a stable job, and a Thursday routine produce, and the arrangement is delivering exactly what it is designed to deliver. It is not evidence about the underlying bond. It is a readout of an experimental configuration.

Change the configuration and you get the other description back. Not as a memory. As a present-tense fact.

The Uncomfortable Implication

I have to be honest about the cost of that, because the standard advice here is dishonest by omission.

You will read, in a great many books, that couples should introduce novelty to keep passion alive. This is true and it is stated as though it were free. It is not free.

The conditions that produce passionate love include uncertainty and obstacle. Some of what made the stairwell is that Priya did not know, in March of that year, whether Tom would still be there in June. The not-knowing was doing work. It was unpleasant work, and she would not go back to it, and it was doing work.

You cannot manufacture genuine uncertainty in a secure relationship without doing damage. Anyone who advises you to create insecurity to reignite desire is recommending you break a thing to hear it ring.

What you *can* manufacture, honestly, is the rest of the list: novelty, arousal, effort, obstacle, and separation.

Novelty is straightforward and undersold: activities neither of you has done, in places neither of you knows, at a difficulty level that involves the real possibility of doing badly. The last clause matters. Aron's finding involves novel *and arousing*, not novel and pleasant. A new restaurant does approximately nothing. A dance class where you are both incompetent in front of strangers does something.

Separation is the one people resist and it is the most reliable. Desire, as Esther Perel has argued at length, requires distance to cross. A partner who is continuously present, continuously available, and continuously known offers nothing to reach for. This is why the return from a work trip is what it is, and why long-distance couples routinely report a passionate register that cohabiting couples envy. It is also why the autonomy side of Chapter Eight's tradeoff is not merely a concession to the more independent partner. It is a functional input to desire.

Watching your partner in their own competence. Perel's observation, and it is well supported by clinical report if not by controlled trials, is that people frequently report renewed attraction when they observe their partner absorbed in something they are good at, among other people, in a context where the observer is not needed. The mechanism appears to be a temporary restoration of otherness: for a moment, the person is not your partner, they are a separate individual with a life you are not inside of.

Priya saw Tom teach a workshop last spring. She has mentioned it four times since.

The Dial, Not The Switch

Return to the duality relations, because this is where the physics earns its place.

The Englert relation says you do not have to pick. Partial interference visibility and partial which-path information can coexist, trading off along a continuous curve. Most real experiments live in the middle.

Most real relationships do too, and the mistake is thinking you are supposed to pick a setting and stay there.

A functional long partnership is not permanently companionate with occasional passionate episodes, and it is not a struggle to maintain the early-stage state. It oscillates deliberately. Long periods of the safe, predictable, deeply unglamorous arrangement that makes a life possible, punctuated by intentional excursions into the other configuration, and both of you knowing that this is what you are doing.

The couples who suffer are the ones who believe the oscillation is a signal. Down into the companionate stretch and they think it is dying. Up into a passionate stretch and they think it is fixed forever. Neither reading is right, because you cannot infer the state of the bond from which description is currently showing.

You can only infer which experiment you have been running.

Practice: The Configuration Audit

Take the last three months and list every substantive thing the two of you did together. Everything, including the dull items.

Sort each into one of two columns.

Familiar and safe. Anything you have done before in roughly the same way. Meals at home. The usual restaurant. Television. Errands. Visiting the same people.

Novel or arousing. Anything genuinely new, physically activating, mildly frightening, involving strangers, or containing a real possibility of failure.

Most couples find the second column has zero or one item across an entire quarter, and then are surprised that they are getting a companionate readout.

Then, the second half. Schedule two things for the next three months that qualify for column two. Real criteria: neither of you has done it, you might be bad at it, and it involves other people. Book them. Pay for them if payment increases the chance of showing up, because it does.

What usually goes wrong: couples choose column-two activities that are actually column one with a new label. A different restaurant is not novelty. A holiday to a place you have been is not novelty. The test is whether you feel any reluctance when you imagine it. Reluctance is the signal that arousal is present. If it sounds nice, it will not work.

The second failure: treating this as an obligation to feel something. You will not necessarily feel the stairwell at the dance class. What the research supports is an effect on relationship quality over time, not a guaranteed emotional event on a given evening. Do the thing, notice what happens over two months, and do not audit yourself on the night.

What Priya Concluded

She and Tom did a two-day sailing course in October, which was Tom's idea and which Priya resented for eleven days beforehand.

She was bad at it. He was worse. There was a forty-minute stretch involving a rope and a rising wind during which both of them were frightened and neither was in charge, and at the end of it he looked at her in a way she recognized from the stairwell, and it had not gone anywhere, it had just not been asked for in eighteen months.

She had thought she was measuring the state of her marriage. She was measuring her calendar.

Interlude: Schrödinger's Text Message

We have been serious for nine chapters. Let us be silly for one, on the grounds that the physics is genuinely funny and nobody ever tells you so.

What follows is a small field guide to the quantum mechanics of ordinary romantic life. Some of it is a stretch. All of the physics is real, including the parts that sound invented, and I have flagged nothing, because you have earned the right to just enjoy it.

Schrödinger's Text Message

You sent it at 9:14. It was read at 9:15. It is now 11:40.

The message currently exists in a superposition of two states. In one, he is driving, or asleep, or in the shower, or one of the eleven thousand ordinary situations that account for essentially every unanswered message in human history. In the other, he read it, felt something complicated, and is composing a reply that begins with the word "look."

Both states are live. Neither is real yet. And the truly quantum feature of this situation is that **you cannot observe it without collapsing it.** Any attempt to check produces a second message, and the second message changes the system you were trying to measure.

Schrödinger invented his cat to make fun of exactly this. He thought a theory that let a cat be alive and dead until someone opened the box was self-evidently absurd, and he wrote the thought experiment as a joke at the theory's expense. It became the most famous image in physics anyway, because everybody loved the cat and nobody read the argument.

The read receipt is the box. Do not double-text at 11:41. That is opening the box out of nerves, and you will not like what the wave function does under those conditions.

The Quantum Zeno Effect, Or Why Watching It Makes It Not Happen

This one is real, has a name, and will improve your life.

In 1977 Baidyanath Misra and George Sudarshan showed that a quantum system which is measured continuously cannot evolve. Every measurement resets it to the state you just found it in. Observe an unstable particle often enough and it will not decay. Watching genuinely does freeze the pot.

They named it after Zeno of Elea, who argued that an arrow in flight is motionless at every instant, and therefore never moves. The physicists gave the paradox a laboratory, and it has since been demonstrated in real systems.

Now think about the last time you were waiting to see whether something was healing.

You checked on Monday. You checked on Wednesday. Every check took the form of a slightly loaded question about how things felt, and every one of those questions required your partner to look directly at a thing that only mends when nobody is looking at it, and so the answer was always "fine, I think," delivered in the voice of somebody who has just been asked to describe a bruise.

Some things in a relationship recover on their own, quietly, in the background, over about three weeks. The single most effective intervention available to you is to stop taking the measurement.

Zeno's arrow only stalls if you keep insisting on knowing where it is.

The No-Cloning Theorem, Or Why You Cannot Date Him Again With Someone Else

There is a theorem, proved by Wootters and Zurek in 1982, stating that it is impossible to make an identical copy of an unknown quantum state. Not difficult. Impossible, in the way that dividing by zero is impossible.

I think about this whenever somebody describes their type.

You do not have a type. You have one relationship that worked, and you have extracted a set of surface features from it, and you are now attempting to reconstruct an unknown state from an incomplete description. This does not work in quantum information theory and it works even less well in dating.

The state you are trying to copy included things you never measured and could not name. It included his particular timing, and a specific year in both your lives, and a version of yourself that no longer exists. What you can specify, tall and funny and works in medicine, is the part that carried the least information.

You cannot clone him. You can only prepare a new state, which is a different and better project, and which occasionally produces something neither of you would have thought to specify.

Bertlmann's Socks

John Bell, of Bell's theorem, opened one of his papers with a story about a colleague named Reinhold Bertlmann, who always wore socks of two different colors.

Nobody ever knew which color would be on which foot. But the moment you saw that the first sock was pink, you knew instantly, with certainty, at any distance, that the second one was not pink.

Bell's point was that this is not mysterious in the slightest, and that people who describe entanglement this way have described nothing. Of course the socks were already different when he put them on. That is the boring explanation, it is completely adequate for socks, and the whole content of Bell's theorem is that it is *not* adequate for particles.

I include this because it is the correct response to anybody, including the author of this book in a weak moment, who tells you that your bond with someone is quantum because you both texted at the same time.

Sometimes you both texted at the same time because you have known each other for nine years and it was raining.

The Casimir Effect, Or Attraction By Absence

Take two uncharged metal plates and put them extremely close together in a vacuum. They will be pushed toward each other.

Nothing is pulling them. What happens is that empty space is not empty; it is full of fluctuating fields, and the gap between the plates is too narrow to accommodate all the possibilities that exist outside the gap. There is more going on outside than inside, and the imbalance presses the plates together.

Hendrik Casimir predicted this in 1948 and it has been measured.

I find it a lovely thing to know. Two objects, drawn together not by any force between them, but by everything that has been excluded from the space they share.

That is what a private world is. Not a bond, exactly. A place with fewer possibilities in it than the world outside, which is precisely why it holds.

Heisenberg Gets Pulled Over

He is doing ninety-five on the interstate. The officer walks up and says, do you have any idea how fast you were going.

Heisenberg says no, but I know exactly where I am.

This joke has been told at every physics department on Earth and I am contractually obliged to include it. Its relationship application is that you have now had this exact conversation about the evening, in which one of you knew precisely where you were and the other knew precisely how fast things were going, and neither of you could produce both at once.

Vacuum Fluctuations, Or Nothing Is Never Happening

Empty space is not empty. At the quantum level the vacuum seethes: fields fluctuate, pairs of particles flicker into existence and annihilate, and the whole thing has an energy that will not go away no matter how much you take out. You cannot get to nothing. Nothing is not available.

So when somebody says nothing is going on between us, they are making a claim that the universe does not support anywhere, at any scale.

Something is always going on. It is just below the threshold of your current instrument. This is the entire premise of Chapter Three and I have snuck it in here with better jokes.

Two Slits, One Dinner Party

You have watched this happen and never had a name for it.

Your partner among his oldest friends is a demonstrably different person than your partner at his mother's house, who is a different person again alone with you at eleven at night. Not performing. Different. The warmth arrives at a different temperature and the jokes have a different shape.

For years you assumed one of these was the real one and the others were versions, and you spent a certain amount of energy trying to determine which.

There is no fact of the matter. Which apparatus, which result. The man who exists at that specific kitchen table with those specific three people is not a costume worn over the man who exists with you. He is a genuine outcome of a genuine arrangement, and so is yours, and you do not get to see all of them at once any more than you get interference and which-path in the same run.

You get one experiment at a time. The consolation is that yours is the one where he sits down.

And One Piece Of Actual Advice

There is an old line among physicists, usually attributed to Feynman in one form or another, that if you think you understand quantum mechanics, you do not understand quantum mechanics.

Feynman meant it as a warning about intuition. I have come to think of it as a description of being married for a long time.

You will not finish understanding this person. There is no year in which the model is complete and you can stop revising. Every settled fact will need updating, and some of your best explanations will turn out to have been convenient rather than true, and the whole thing will keep producing results you did not predict.

The physicists did not find this depressing. They found it the most interesting thing that had ever happened to them, and they spent their entire lives on it, and they were happy.

Right. Back to work.

PART FOUR

Change: What Moves

Chapter Ten: Tunneling, Or Getting Through The Wall You Cannot Climb

The sun should not be burning.

This is not a figure of speech. Take the standard model of stellar fusion, calculate the temperature at the core of the sun, and work out how much kinetic energy two protons have when they collide there. Then calculate how much energy they would need to overcome their mutual electrical repulsion and get close enough for the strong nuclear force to bind them.

The protons do not have enough. They are short by a wide margin. At fifteen million degrees, the overwhelming majority of protons in the sun's core are moving far too slowly to climb the electrostatic barrier between them. Classically, fusion in the sun is impossible, and the sun should be a cold ball of hydrogen, and you should not exist.

In 1928 George Gamow worked out why it happens anyway, and the answer is the subject of this chapter.

The Wall That Is Not Solid

A quantum particle is described by a wave function, and a wave function does not stop abruptly at a barrier.

When a wave meets a wall it does not simply bounce. Some of it penetrates, and inside the barrier the amplitude decays, dropping off exponentially with depth. If the barrier is thin enough, the amplitude has not decayed to nothing by the time it reaches the far side, and a small but nonzero portion of the wave emerges, still travelling.

Which means there is a real probability of finding the particle on the other side of a barrier it never had the energy to cross.

It did not break through. It did not tire the wall out. It did not gather itself and try harder. The particle's existence was never confined to one side in the first place; the wave extended through the barrier all along, and what we call tunneling is simply the fact that a measurement sometimes finds it over there.

This is not exotic. It is why the sun burns. It is how radioactive alpha decay works, which was Gamow's original application. It is the operating principle of the scanning tunneling microscope, which images individual atoms by measuring a tunneling current across a gap. It is a design constraint in every modern processor, because when transistor gates get thin enough, electrons tunnel across them and leak, and managing that leakage is one of the central problems of chip design.

The Two Numbers That Govern It

Here is the part that carries the whole chapter, and it is quantitative.

The probability of tunneling depends on three things: the height of the barrier, its width, and the mass of the particle. And the dependence is exponential.

Width matters enormously. Because the amplitude decays exponentially inside the barrier, doubling the width does not halve the probability; it can reduce it by many orders of magnitude. In a scanning tunneling microscope, changing the tip-to-surface gap by a single atomic diameter changes the current by roughly a factor of ten. That extreme sensitivity to width is what makes the instrument work.

Energy matters much less than you would expect. Raising the particle's energy helps, but sub-exponentially compared to thinning the barrier. If you want to get through a wall, making yourself faster is a far worse strategy than making the wall thinner.

Mass matters, and lighter is better. This is why electrons tunnel readily, protons less so, and why a baseball has never tunneled through a bat. The probability for a macroscopic object is not zero; it is a number so small that the age of the universe is not a meaningful amount of time to wait.

And then the fourth factor, which is not in the barrier at all.

Attempt frequency. In the sun, the probability that any given proton pair fuses on any given approach is astronomically small. The number of proton pairs is also astronomical, and each approach happens on a timescale of about ten to the minus twenty-one seconds. Multiply a vanishing probability by an enormous number of attempts and you get a star.

Low odds per attempt, plus an absurd number of attempts, equals certainty.

The Wall Nadia And Wes Could Not Climb

By year nine they had a subject they did not discuss.

It was the third year, and what happened in it, and specifically a period of about four months during which Wes was somewhere else emotionally and Nadia knew it and neither of them ever said so. Nothing that would show up on a list of betrayals. It was just that there had been a woman at his firm, and nothing happened, and Wes had thought about it, and Nadia had known he was thinking about it, and both of them had decided, separately and without discussion, that the survivable option was silence.

Six years later the silence had become architecture. It had a shape you could walk around. Certain conversations bent to avoid it. Neither of them could have raised it, because raising it would require crossing six years of accumulated implication all at once, and the energy required had grown every year, and neither of them had it.

That is a barrier. Not a disagreement. A disagreement has terms. This had no terms, and every year of not discussing it made it taller.

Here is what did not work, and they tried all of it: waiting for the right moment, which never arrives, because the right moment for a hard conversation is a category error. Deciding to have a serious talk, which raised the stakes and therefore the barrier. Couples' weekends, during which both of them were aware of the subject the entire time and neither raised it and both came home defeated.

Every one of those is an attempt to add energy. To take a run at the wall.

Thin The Barrier

What worked was so much smaller than any of that.

Wes said, in a car, at a red light, apropos of a podcast about something unrelated: "There was a stretch in the third year when I was pretty checked out. I have never known how to say that."

Nadia said, "I know."

And then the light changed and they drove for two minutes without speaking and then talked about it for the rest of the drive and, in pieces, for the following two months.

Look at what that moment had going for it. A car, so no eye contact, which lowers threat. A red light, which imposed a time limit and therefore capped the stakes. An unrelated conversational context, so no announcement and no run-up. Eleven words, which is a small enough commitment that retreat remained available. And no request for a response, so Nadia was not put on the spot.

Every one of those features made the barrier thinner. Not one of them made Wes stronger.

This is the central practical claim of the chapter and it runs against every instinct. **You do not get through a stuck place by mounting a bigger effort. You get through it by reducing the width of the thing you are trying to cross.**

The specific reductions that work:

Reduce the size of the disclosure. Not the whole subject. One sentence about one part of it. The subject is six years wide; the sentence is eleven words.

Reduce the demand on the other person. A statement that does not require a response is far easier to make and far easier to receive. Requests raise the barrier for both parties.

Reduce the formality. The scheduled serious conversation, with both people seated facing each other and the phones away, is the highest-barrier configuration available, which is unfortunate given that it is what everyone recommends. It works well for logistics and badly for the things that have never been said. Cars, walks, kitchens at eleven at night, and any activity involving parallel attention are lower.

Reduce the stakes of failure. The reason people cannot raise the hard thing is usually not that the conversation would be painful. It is that the conversation might be catastrophic and irreversible. Anything that makes retreat available lowers the barrier: "I do not want to get into all of it tonight" spoken in advance is a permission slip that makes the attempt possible.

Attempt Frequency

Now the other half, which is where the sun comes in.

John Gottman's observational research at what he called the Love Lab produced a finding that I think is the single most useful empirical result in the field, and it is about frequency rather than magnitude.

Gottman and colleagues coded what he calls bids for connection: small, low-stakes attempts to engage a partner's attention. A comment about a bird outside. A hand on a shoulder. Reading something aloud. Dozens per day in an ordinary household, almost all trivial.

Each bid can be met with turning toward, turning away, or turning against. And the rate of turning toward, in Gottman's observational samples, differed dramatically between couples who remained married and couples who later divorced. The often-quoted figures are around eighty-six percent versus thirty-three percent.

I want to be careful with this. It is observational, the samples were not large, and Gottman's predictive accuracy claims have drawn substantial methodological criticism, particularly from Richard Heyman and others, on the grounds that models fitted and tested on the same small samples will overstate their power. Treat the specific percentages as illustrative rather than as constants of nature.

What survives the criticism, and is corroborated across multiple labs, is the qualitative structure: the health of a relationship lives in the accumulated response to a very large number of very small events, not in the handling of a small number of large ones.

That is attempt frequency. Each bid has a low probability of producing anything. There are thousands of them. The stuck things in a relationship do not give way to one heroic push; they give way somewhere around bid four thousand, in a car, at a red light.

Sudden Gains

There is a related finding in psychotherapy research that deserves more attention than it gets.

Tony Tang and Robert DeRubeis, in 1999 and in subsequent work, documented what they called sudden gains: discrete sessions in cognitive therapy for depression after which a patient's symptoms dropped sharply and stayed down. These are not gradual improvements. In their samples, a substantial minority of patients showed a single between-session interval accounting for a large share of their total improvement, and those patients had better outcomes at follow-up.

The pattern has been replicated across therapy types and disorders, though there is real debate about mechanism and about how much of it is measurement artifact.

What interests me is the phenomenology, which matches tunneling far better than it matches climbing. Change in a stuck system is often not proportional to effort. Long periods of apparently nothing, then a discontinuous jump. And critically, in Tang and DeRubeis's analysis, the sessions preceding a sudden gain did not look special. You could not tell it was coming.

This has a practical consequence that I think matters enormously for people in long stuck relationships: **the absence of visible progress is not evidence that the attempts are not working.** In a tunneling process, every attempt looks identical to every failed attempt, right up until one does not.

Where This Advice Is Dangerous

I have to put the guardrail in, because "keep trying, breakthroughs come without warning" is exactly the sentence that keeps people in situations they should leave.

Tunneling is not a general argument for persistence. It is an argument about a specific structure: a barrier that is genuinely finite, with a system on both sides that would be stable if it got through.

That is not every wall. Some walls are not barriers between two states of the same relationship; they are the edge of the relationship. The distinctions I would use:

Is the barrier maintained or inherited? Nadia and Wes's silence was inherited. Nobody was rebuilding it deliberately. A barrier that your partner actively reinforces every time you approach it is not the same object, and thinning it is not something you can do alone.

Is there anything on the other side? Tunneling gets you through to a state. If you cannot describe what you would find, you may be pushing at a wall for its own sake.

Is the cost of attempts rising? In the physics, attempts are free. In a relationship, they are not, and if each attempt is met with contempt or punishment, then the attempt frequency strategy is not a path to a breakthrough. It is a description of how people get worn down.

And if there is any dimension of fear in it, the frame does not apply at all. This chapter is about stuck conversations between people who are safe with each other. It is not about anything else, and no metaphor from physics should ever be used to encourage someone to keep trying with a person who frightens them.

Practice: The Eleven Words

Identify one thing you have not said to your partner that you would like to have said. Not the hardest one. Something in the middle.

Write the full version, everything you would want to convey. It will run to a page. Then cut it to a single sentence that contains one true thing and no accusation and no request. Cut it again until it is under fifteen words.

Then wait for a low-barrier context. Not a scheduled talk. A car, a walk, a kitchen, a moment of parallel attention. Say the sentence. Do not follow it with anything.

Then let there be silence, which will feel intolerable for approximately six seconds.

What usually goes wrong: people cannot resist adding a second sentence that explains, softens, or requests, and the addition restores the entire barrier. The power of the eleven words is that they demand nothing. As soon as you add "and I need you to tell me how you feel about that," you have handed the other person a task, and tasks are heavy.

The second failure: waiting for the right context turns into waiting forever. Give yourself two weeks. If no low-barrier moment has occurred by then, you are avoiding, and the honest move is to manufacture one. Suggest a drive. There does not have to be a destination.

What Gamow Understood

The thing I find beautiful about the tunneling calculation is that it did not discover a new force. It did not find some hidden extra energy that the protons secretly had.

It found that the question had been wrong. Physicists had been asking whether the protons could get over the barrier, and the answer was no, and the answer remained no, and the sun burned anyway, because getting over was never what was happening.

Six years of silence sat between Nadia and Wes like a wall neither could climb, and neither of them ever climbed it. He said eleven words at a red light, and it turned out that some part of him had been on the other side the whole time.

Chapter Eleven: Many Worlds, Or The Life You Did Not Choose

There is a man in Lisbon that Dev did not move to be with in 2019.

Not Marcus. Someone else, before Marcus, when Dev was thirty-one and had an offer and a lease he could have broken and a father who was ill in a way that turned out to take four more years. He stayed. It was a defensible decision made for good reasons, and he has never fully stopped running it.

He does not think about the man much. What he thinks about is the version of himself who went. That version has, in Dev's imagination, a specific apartment with light coming in from a particular angle, and a facility with Portuguese that the actual Dev has never acquired, and a relationship to his own life that the actual Dev suspects is more relaxed.

The other Dev is completely fictional. Dev knows this. It has never helped.

Everett's Proposal

In 1957 a graduate student named Hugh Everett III submitted a doctoral thesis at Princeton that proposed the most economical and most alarming solution to the measurement problem yet offered.

Recall the problem. Quantum mechanics describes systems evolving smoothly into combinations of possibilities. Experiments deliver single definite outcomes. The standard treatment patched this by adding a second rule: at measurement, the state collapses to one outcome, chosen randomly with probabilities given by the mathematics.

Everett's move was to delete the second rule.

His proposal, which he called the relative state formulation and which was later popularized as the many-worlds interpretation, is that nothing collapses. The wave function evolves smoothly, always, without exception, including through measurement. What happens when you measure a particle in superposition is that *you* become entangled with it. The combined system, you plus the particle, is now in a superposition of you-having-seen-up and you-having-seen-down.

Both branches are equally real. Each contains a version of you who saw one definite result and has no access to the other. Nothing is selected. The appearance of a single outcome is what being one branch feels like from the inside.

The theory is, in a specific technical sense, the simplest one available: it has fewer rules than the alternatives. It is also the one that requires you to believe in an unimaginable multiplicity of unobservable worlds, which most people find to be a steep price for elegance.

The Comfort Does Not Work

Suppose many worlds is true. Suppose there is a branch in which Dev moved to Lisbon in 2019.

That branch does not contain him. It contains a different person, who diverged seven years ago, who has had seven years of experiences Dev has not had, and who is by now no more Dev than a sibling would be. The two share a history up to a point and nothing after it.

More to the point: the Lisbon branch, if it exists, is causally inaccessible. Everett's branches do not interact after decoherence separates them. There is no transfer of information, no consolation, no benefit of any kind flowing from that world to this one.

And most decisively: the imagined Lisbon Dev is not a branch of the wave function anyway. He is a construction, assembled by a mind that is very good at building coherent narratives and very bad at simulating the parts it does not want to include. The imaginary version has the light and the Portuguese and does not have the fights, the boredom, the illness that would have found him anyway, or the specific ordinary disappointments that any actual seven years contains.

That asymmetry is not a personal failing. It is a well-documented feature of how people forecast.

What The Research Says About The Road Not Taken

Affective forecasting is systematically wrong. Daniel Gilbert, Timothy Wilson, and colleagues have documented across many studies that people substantially overestimate both the intensity and the duration of their emotional responses to future events. They call it the impact bias. The mechanism they propose is that we simulate the event in isolation, without the enormous quantity of ordinary life that will surround it, and without accounting for what they call the psychological immune system, the largely unconscious set of processes that rationalize, reframe, and adapt.

The consequence for counterfactual worlds is direct. When Dev imagines Lisbon, he simulates the light and omits the adaptation. He is comparing a curated highlight to a lived reality, which is not a comparison at all.

Counterfactual direction determines suffering more than outcome does. In 1995 Victoria Medvec, Scott Madey, and Thomas Gilovich published a study of Olympic medalists that has become deservedly famous. They coded athletes' facial expressions and interviews and found that bronze medalists appeared happier than silver medalists.

The explanation is counterfactual comparison. The silver medalist's most available alternative is gold, which was one place away. The bronze medalist's most available alternative is fourth place, and no medal at all. The objective outcome is better for silver. The comparison is worse.

You choose your comparison class, usually without noticing, and it determines how the same life feels.

Maximizing costs you. Barry Schwartz and colleagues distinguish maximizers, who search for the best available option, from satisficers, who search for one that meets their standard and then stop. Maximizers, across several studies, achieve objectively better outcomes on some measures, such as starting salary in a job search, and report lower satisfaction, more regret, and more counterfactual thinking. Later research has complicated the picture, with some evidence that the effect is driven mainly by the tendency to ruminate about alternatives rather than by high standards themselves, which is a useful refinement rather than a refutation.

Applied to relationships, the maximizer's question is whether someone better exists. In a population of billions, the answer is trivially yes, and it is also completely useless, and the question functions only as a mechanism for generating dissatisfaction with whoever is actually in the room.

Believing in "the one" makes things worse. Raymond Knee and colleagues have studied what they call implicit theories of relationships, distinguishing destiny beliefs, the view that partners are either compatible or not, from growth beliefs, the view that relationships develop through effort. Destiny believers respond to relationship difficulty as diagnostic evidence of mismatch and are more likely to disengage. Growth believers respond to the same difficulty as a problem to work on. The pattern holds across several studies, with destiny beliefs predicting shorter relationships and stronger reactions to initial dissatisfaction.

If there is one branch of the multiverse containing your correct person, then every difficulty is evidence that you are in the wrong branch. That is not a romantic idea. It is a mechanism for leaving.

Sliding Versus Deciding

I want to introduce one more line of research here, because it is the most actionable material in the chapter and it is not widely known outside the field.

Scott Stanley, Galena Rhoades, and Howard Markman have written extensively about what they call sliding versus deciding. Their observation is that a very large proportion of major relationship transitions, moving in together, becoming sexually exclusive, having a child, marrying, happen not through a decision but through a gradual accumulation of circumstance that makes the transition the path of least resistance.

Their concern is with what they call inertia: transitions that increase the cost of leaving, entered without a deliberate commitment, can result in people remaining in relationships they would not have chosen. Their research on cohabitation before engagement found associations with lower marital quality and higher divorce risk, with the deliberateness of the transition appearing to matter.

Here is why this belongs in the many-worlds chapter.

The counterfactual that actually torments people is almost never the one about the different partner. It is the one about *having chosen at all*. Dev's regret is not really about the man in Lisbon. It is about the fact that he cannot locate a moment when he decided to stay. He remembers a period during which the decision made itself, and the absence of a moment of agency is what has kept the branch alive in his imagination for seven years.

A decision that was actually made can be owned, defended, and eventually stopped relitigating. A slide cannot, because there is nothing to own.

Which suggests the intervention, and it is a strange one. **You can decide retroactively.** Not change the past; take possession of it. Dev can, at thirty-eight, look at the choice he slid through at thirty-one and say, explicitly and out loud, *I chose to stay for my father and I* *would choose it again*. Whether or not that is what happened at the time, saying it converts a slide into a decision, and decisions close branches.

That is not self-deception. The facts are unchanged. What changes is whether he is the author of his life or a person to whom his life occurred.

The Branch You Are In

Here is the reframe I actually want to offer, and it is closer to the physics than the usual one.

The self-help version of many worlds says: all your choices happen, so relax. That is false comfort, because the other branches are inaccessible and imaginary.

The version I find defensible is the opposite. **In many worlds, this branch is not one of many equivalent options. It is the only one you are in, and it is completely real, and no amount of possibility elsewhere provides any support to it whatsoever.**

Everett's branches do not help each other. There is no fund of unlived lives you can draw on. Every branch has to be inhabited by the version of you that is in it, working with what that branch contains.

Which means the multiplicity, if it is real, has exactly one practical implication: the alternatives are not available to you. Not because they do not exist, but because you are not the person in them and cannot become that person. The only life you have any influence over is this one, and the imaginary Lisbon apartment with the good light is competing for resources that could be spent on the actual apartment you actually live in.

The multiverse, properly understood, is an argument against wistfulness rather than for it.

Practice: Closing The Branch

This is a written exercise, done once, and it takes an hour. It is uncomfortable and it works.

Step one. Write down the counterfactual you return to. Be specific about what you imagine. The place, the person, the version of yourself.

Step two. Now populate it honestly. Write out the mundane content that your imagination has been omitting. Where does the money come from. What is the fight you would be having by year three, given who you are, since you would have brought yourself. What happened to your father in that branch, and were you there. What is dull about that life.

Most people find this genuinely difficult, and the difficulty is the finding. The counterfactual has survived this long precisely because it has never been fully specified.

Step three. Write the decision you actually made, in the first person and active voice, as a decision. Not "it did not work out" or "the timing was wrong" or "I could not have gone." *I chose.* Then the reasons, and they should be reasons you can stand behind.

Step four. Write one sentence about what this branch has that the other one does not. Not a better thing. A specific thing, that exists, that you would not have.

What usually goes wrong: people write step three as a justification, defending the choice against an imagined accuser, which keeps the counterfactual alive by continuing the argument. The point is not to win. It is to take authorship. There is a tonal difference between "I had no choice, my father was sick" and "I chose my father," and the second one closes something the first one does not.

The other failure: doing this about a decision that is still open. This exercise is for closed branches. If you are currently deciding whether to leave someone, do not use it to prematurely foreclose the question. That is a different problem and it needs a different chapter, which is the next one.

What Dev Says Now

He still has the apartment in his head. He suspects he always will, and I think that is normal and not a defect.

What changed is that he stopped treating it as a report from a real place. He describes it now as a piece of fiction he wrote when he was thirty-one and frightened, which he keeps around because it tells him something about what he wanted at the time and nothing at all about what was available.

And when Marcus asked him, in the second year of six time zones, whether he was going to move to Lisbon, Dev said yes, and gave a date, and it is on a calendar.

That is the thing about branches. You cannot visit the ones you did not take. You can still notice that you are standing at another one, and that this time you get to be the person who decides rather than the person who slid.

Chapter Twelve: Entropy, Or Why Love Requires Continuous Work

I am going to break my own rules for one chapter.

Everything else in this book comes from quantum mechanics. This chapter comes from thermodynamics, which is a different field, was developed earlier, and concerns steam engines rather than electrons. It does not belong in a book called The Quantum Physics of Relationships, and I have put it here anyway, because it is the most useful physics for anyone trying to keep a relationship alive and leaving it out would be a dereliction.

I am telling you this rather than quietly blurring the categories, which is what the first edition did.

The Law With No Exceptions

The second law of thermodynamics states that the entropy of an isolated system does not decrease over time.

Entropy is often glossed as disorder, which is close enough to be useful and imprecise enough to cause trouble. The more careful statement is that entropy counts the number of microscopic arrangements consistent with a given macroscopic description. A shuffled deck has high entropy not because shuffling is chaotic but because there are astronomically many shuffled orders and exactly one sorted one. Systems drift toward high-entropy states for the same reason a coin flipped a thousand times lands near five hundred heads: there are vastly more ways to be disordered than ordered, and nothing is steering.

The second law has a claim to being the most secure principle in all of science. Arthur Eddington's remark, that if your theory is found to be against the second law there is nothing for it but to collapse in deepest humiliation, is not hyperbole. Individual laws of mechanics can be revised. This one has survived every revision underneath it.

Two consequences matter here.

Order requires energy, continuously. A refrigerator keeps its interior cold, which is a local decrease in entropy, by consuming electricity and dumping more entropy into your kitchen than it removed from the food. The total always goes up. Local order is always purchased, never free, and the purchase never ends. Unplug the refrigerator and it does not stay cold out of momentum.

The arrow only points one way. Nearly every fundamental law of physics is time-symmetric; run the equations backward and they work. The second law is the exception, and it is the reason time has a direction at all. An egg does not unbreak. A cup of coffee does not spontaneously separate into hot and cold halves. Not because it is forbidden by mechanics but because the number of arrangements that look like an unbroken egg is so vanishingly small compared to the number that look like a mess.

What Erodes

With that stated, here is what I think the frame captures that nothing else in this book does.

There are vastly more ways for a relationship to be disorganized than organized, and nothing is steering.

Consider what a functioning partnership requires simultaneously. An accurate and current model of the other person. A stable division of labor. A shared account of your history. A working repair process. Aligned expectations about money, time, sex, and family. A private vocabulary. Regular unstructured contact. And a set of agreements each of you can predict from the other.

Every one of those is a specific, improbable configuration. Every one of them has a thousand ways to be slightly wrong and one way to be right.

And here is the thing about the drift: **it requires no cause.** This is the piece people cannot accept, and it is why the question "what went wrong" so often has no answer. Nothing went wrong. Wes's model of Nadia was accurate in 2019 and was not updated, and she changed, as people do, and now the model is wrong. No event. No fault. The model did not become wrong through any process; it became wrong through the absence of a process.

You can see the same thing in every component. The division of labor was fair when they set it and then their jobs changed. The private vocabulary was rich and then stopped acquiring terms. The repair process worked and then one of them got busy and skipped a repair, and skipping was easier the second time.

The second law does not say a saboteur is at work. It says that in the absence of continuous energy input, systems move toward the overwhelmingly more numerous disordered states, because there are more of them.

The Asymmetry

The arrow of time gives us the second thing, and it is the harsher one.

Building and destroying are not symmetric operations, and they do not take comparable amounts of effort.

Roy Baumeister and colleagues made this general case in a widely cited 2001 review titled "Bad Is Stronger Than Good," surveying evidence across many domains. Negative events have stronger and more lasting effects than positive events of comparable magnitude. Bad information is processed more thoroughly. Negative impressions form faster and are more resistant to disconfirmation. Bad feedback affects behavior more than good feedback.

In relationships specifically, this shows up as the well-known asymmetry between damage and repair. A single contemptuous sentence, delivered in eight seconds, can require months of consistent behavior to offset. This is the origin of the various positive-to-negative ratios that circulate in the popular literature; whatever the correct number is, and it is probably not a constant, the direction is not in dispute.

Here is why this is thermodynamic rather than merely sad. The sentence is a low-effort, high-entropy operation. It requires no preparation, no coordination, no sustained attention. Undoing it requires the opposite: sustained, coordinated, attentive behavior over an extended period, all of which is energy input against the gradient.

You can smash a glass by dropping it. Reassembling it is not the same task performed backward.

The practical implication is not "never say anything harsh," which is neither possible nor desirable. It is that the accounting is not one to one, and people persistently act as though it is. They think an apology settles the balance because it addresses the same event. It does not, because the apology is one operation and the damage was a state change.

What Gets Imported

Now the useful half, and it belongs to a Russian-born chemist named Ilya Prigogine, who won the 1977 Nobel Prize for working out how systems escape the gloomy reading of the second law.

His answer was that the law applies to isolated systems, and almost nothing interesting is one. Open systems, fed by a continuous flow of energy, can maintain and even build internal order indefinitely, exporting their disorder somewhere else. He called them dissipative structures. Every living thing is one. A candle flame is one. So is a marriage.

Open systems can maintain and increase order indefinitely as long as energy flows through them. A living cell has maintained an exquisitely improbable internal organization for its entire existence. It does this by eating.

The relational question is therefore not how to prevent entropy, which is not available, but what your system imports and how reliably.

Attention is the primary input. Not affection, not goodwill, not love in the abstract. Attention, measured in units of time during which nothing else is happening. This is the energy source, and every other maintenance process runs on it. The couples who fail are almost never the couples who stopped loving each other; they are the couples whose attention was fully consumed by other systems that also required energy, such as children and employers, both of which are extremely efficient at extracting it.

Novelty is a second input, for the reasons in Chapter Nine, and it functions differently: it does not maintain the existing order, it adds new structure.

External inputs matter and are undercounted. Friendships, community, extended family, time apart, work that is going well. Couples who import nothing from outside are running a closed system, and closed systems have only one trajectory. This is the mechanism behind the observation that isolated couples do badly. It is not mysterious. They have no energy source.

And there is a maintenance schedule. Every physical system with a maintenance requirement is serviced on a cycle, not when it fails. Nobody changes the oil in their car when the engine seizes. And yet the standard approach to relationships is to intervene at breakdown, which is the most expensive possible time, because by then the disorder has propagated into every subsystem.

Consider what a service schedule would look like. Weekly: the enclosure from Chapter Three, twenty-five minutes of unstructured contact. Monthly: a check on the division of labor, which drifts faster than anything else and which nobody tracks. Quarterly: one column-two activity from Chapter Nine. Annually: the phase audit from Chapter Six and the distribution conversation from Chapter Eight.

That is maybe forty hours a year. It sounds like a lot until you compare it to what people spend on the maintenance of a house, or a car, or a body, none of which they expect to run on affection.

The Objection

There is an objection to this entire chapter that I hear constantly and that I want to take seriously rather than dismiss.

It goes: if it takes this much work, is it even love? Should it not be easier with the right person? Are we not describing a job?

The honest answer has two parts.

The first part: yes, it is work, and the belief that it should not be is a historically recent and culturally specific idea that has done a great deal of damage. Eli Finkel's analysis in The All-or-Nothing Marriage traces how expectations of marriage have climbed from economic partnership through companionship to the current expectation that a spouse should supply self-actualization, and argues that the highest-functioning modern marriages are better than any in history while the average has become harder, precisely because the demands have risen faster than the investment. His prescription is either to invest much more or to expect less, and he is explicit that the middle path, high expectations with low investment, is the one that fails.

The second part, and this is the part I believe more strongly: the work is not separate from the love. This is where the refrigerator metaphor misleads, and I should say so.

A refrigerator's compressor is not part of the coldness. It is a mechanism that produces coldness. But the twenty-five minutes of unstructured talk is not a mechanism that produces intimacy; it *is* the intimacy. The attention is not the cost of the relationship. It is the relationship. There is no additional thing being purchased.

Which means the objection contains a category error. Asking whether love should require this much work is like asking whether a fire should require this much burning.

Practice: The Maintenance Audit

One evening. Both of you. A piece of paper with four columns.

Column one: what the system requires. List the components. Accurate models of each other. Division of labor. Repair process. Shared account of your history. Unstructured contact. Physical intimacy. Alignment on the big questions. Outside inputs.

Column two: when was it last serviced. For each component, when did you last actually attend to it, not incidentally but on purpose. Most couples find that three or four items have not been touched in years.

Column three: what is drifting. Where is your model out of date. What was fair three years ago and is not now.

Column four: the schedule. For each component, how often it should be serviced, and on what day.

Then put the recurring items on an actual calendar, with alerts, like a dental appointment.

What usually goes wrong: the schedule is set too ambitiously in the first flush of the audit, collapses within six weeks, and the collapse is read as evidence that the whole idea was artificial. Set it lower than you want to. A twenty-five minute weekly window that actually happens fifty times a year beats a two-hour weekly window that happens nine times.

The second failure: one partner takes ownership of the schedule and becomes the maintenance department, which converts the entire project into a fresh source of resentment and reproduces the mental-load problem the audit was supposed to expose. If the calendar invites are coming from one person, the audit has failed, whatever the calendar says.

The Thing Eddington Got Right

Eddington also wrote that the law of entropy holds the supreme position among the laws of nature, and I have always found the phrasing slightly menacing, as though the universe were run by an unsympathetic bureaucrat.

But there is something else in it, which I did not see for a long time.

The second law is the reason anything is ever an achievement. In a universe without it, order would be free and permanent and would mean nothing. The improbability of your relationship, the fact that this specific configuration of two people who understand each other's histories and vocabularies and silences is one arrangement out of an unimaginable number of alternatives, is what makes it worth something.

It is going to erode. It is eroding now, in every hour that nothing is put into it.

That is not a tragedy. That is what makes the putting-in count.

Chapter Thirteen: Symmetry, Or What Fairness Actually Is

In 1918 a mathematician named Emmy Noether proved a theorem that Einstein described as a piece of penetrating mathematical thinking and that most physicists now regard as one of the deepest results in the field.

She was, at the time, working at Göttingen without a salary and without a position, because the university would not appoint a woman to the faculty. David Hilbert, arguing on her behalf, is said to have pointed out that the senate was not a bathhouse. She lectured for four years under his name.

The theorem she proved is this: **every continuous symmetry of a physical system corresponds to a conserved quantity.**

That sentence does not sound like it should change your marriage. Give me a few pages.

What Noether Found

A symmetry, in physics, means that something can be changed without changing the outcome. If an experiment gives the same result in Chicago as in Lisbon, the laws are symmetric under translation in space. If it gives the same result today as last year, they are symmetric under translation in time. If it gives the same result whichever direction the apparatus faces, they are symmetric under rotation.

Noether showed that each of these symmetries is not merely an aesthetic feature. Each one *produces* a conservation law, rigorously and inevitably.

Symmetry in time gives you conservation of energy. Symmetry in space gives you conservation of momentum. Rotational symmetry gives you conservation of angular momentum.

This is a startling result. Conservation of energy had been an empirical rule, observed to hold, without an explanation for why. Noether showed that it is not an independent fact about the universe at all. It is a consequence of the fact that the laws of physics do not change from one day to the next.

Read that again, because it is the entire chapter: **energy is conserved because the rules stay the same over time.** If the laws of physics were different on Thursday, energy would not be conserved. The constancy of the rules is what makes the quantity reliable.

Why Inconsistency Costs More Than Unfairness

Here is a situation that most people find puzzling until they see it this way.

Consider two households. In the first, one partner does seventy percent of the domestic work. This has been true for years, both know it, it was discussed and agreed on for reasons connected to their jobs, and it has not changed.

In the second, the split hovers near fifty-fifty on average, but it varies enormously week to week depending on mood, on who is more stressed, on whether someone is annoyed about something unrelated. Some weeks one person does almost everything. The following week it reverses without discussion.

By any measure of equality, the second household is fairer. And in my experience and in the research, the second household generates far more resentment.

The equity literature gives part of the explanation. Beginning with Elaine Hatfield and colleagues in the 1970s, equity theory found that both under-benefited and over-benefited partners report distress, and that *perceived* fairness predicts satisfaction better than objective equality does. Perception, in turn, depends heavily on whether the arrangement is understood, expected, and legible.

But I think Noether gets closer to the mechanism. In the first household the rules are invariant, so something is conserved: each person can predict the coming week, can plan around it, can trust the arrangement. That conserved quantity is worth more than the ten percent of labor it costs.

In the second household nothing is invariant, so nothing is conserved. Every week is a fresh negotiation conducted through inference and irritation. There is no quantity either person can count on, and the absence of a conserved quantity is experienced as unfairness even when the ledger balances.

Predictability is not the opposite of generosity. It is the precondition for it.

This is why the phrase "we just figure it out as we go" is such a reliable predictor of trouble, and why couples resist hearing it. Flexibility feels like a virtue. It is a virtue on top of an invariant structure, and it is corrosive as a replacement for one.

Trust Is A Conservation Law

Follow the logic one more step.

What people mean by trust, when they examine it, is rarely a belief that their partner is good. It is a prediction that certain things will hold: that what was true last month is true this month, that a stated intention will still be operative tomorrow, that the person who was kind on Monday will not be someone else on Thursday.

Trust is the conserved quantity produced by behavioral invariance. It is not a feeling you decide to have. It is a quantity that accumulates automatically when the rules do not change, and that cannot be manufactured by any other route.

This explains something about repair that people find deeply unfair, and which I have watched wreck reconciliations.

After a serious breach, the person who caused it typically wants to restore trust through intensity: large gestures, extensive discussion, demonstrations of remorse. All of that is energy, and by Noether's logic energy is not the input that produces the quantity in question. Only invariance produces it, and invariance takes time by definition, because a rule that has held for two weeks has held for two weeks and there is no way to compress that.

The frustrated question, "what more do you want me to do," has an answer that is unbearable to hear: nothing more. The same thing, for much longer.

Spontaneous Symmetry Breaking

Now the second half of the physics, and it is the half that explains how two people who started as equals ended up in an arrangement that neither of them chose.

Balance a pencil perfectly on its point. The situation is completely symmetric: no direction is preferred, gravity pulls straight down, the laws governing the pencil have no built-in bias toward north or east.

Let go. The pencil falls in some direction. The symmetry of the laws has not been violated; the *state* has simply selected one outcome from a symmetric set of possibilities, and now the pencil is lying east, and the symmetry is gone.

This is called spontaneous symmetry breaking, and it is one of the central ideas in modern physics. A ferromagnet above its Curie temperature has no preferred direction; cool it below, and the atomic magnetic moments align, and the material now points somewhere, though nothing in the physics specified where. The Higgs mechanism, which gives elementary particles their masses, is a symmetry-breaking story of the same shape.

Two features matter here.

The broken state is lower in energy than the symmetric one. The fallen pencil is more stable than the balanced one. Symmetry is expensive to maintain; breaking it is a relaxation.

Which way it breaks is not determined by the laws. Some tiny fluctuation, some accident, decides. The outcome is arbitrary and the arbitrariness is invisible afterward, because once broken, the state looks entirely natural.

The Roles Nobody Chose

Nadia and Wes, in their first year, were symmetric with respect to almost everything domestic. Both could cook. Neither had strong preferences about the apartment. Both had jobs of similar demand.

Eleven years later Nadia manages every medical appointment, every school communication, every gift, every relationship with both families including his, and the entire mental inventory of what the household needs. Wes handles the cars, the finances, and anything involving a ladder.

Neither of them decided this. There was no conversation. What happened is that during a stretch in the second year Nadia was between jobs and took on the scheduling, and it was efficient, and by the time she was working again the arrangement had a groove in it. Then their first child arrived, and the person who already had the model of the household kept it, because handing over a model is more work than continuing to run it.

Spontaneous symmetry breaking. Nothing in either of them specified this outcome. A small fluctuation, four months of unemployment, selected a direction, and the state relaxed into it, and eleven years later it looks like a fact about who they are.

The sociology here is well documented. Allison Daminger's work on cognitive labor found that the anticipation and monitoring components of household work are substantially more gendered than the execution components, and that couples routinely describe their arrangements in terms of individual preference and aptitude, which she notes functions to make an unequal outcome feel chosen. The specialization also compounds: the partner who tracks a domain gets better at it, which makes handing it over more costly, which entrenches it further.

And it is genuinely low energy. That is the part people miss. The broken state is stable *because* it is efficient. Every attempt to restore symmetry, to actually share the mental inventory, costs more in the short term than continuing does. This is why couples slide back within weeks of deciding to change it.

Restoring Symmetry Requires Energy Input

If you want the pencil upright, you have to lift it and hold it there, and the moment you stop holding it, it falls again.

The practical translation is that redistributing a broken domestic symmetry cannot be accomplished by an agreement. Agreements are free and the broken state is a stable minimum, so the system returns.

What actually works, in the accounts I trust, has two features.

Transfer of the model, not the tasks. Handing someone a task leaves the tracking with the original owner, which leaves the load exactly where it was and adds a management burden on top. The transfer has to be of an entire domain, including the noticing. Not "can you book the dentist," which is a task assigned by the person who remembered. "Dental is yours, including remembering that it exists," which is a domain.

A period of deliberate inefficiency. The receiving partner will do it worse for several months. Appointments will be missed. The original owner will know they are being missed and will have to not intervene, which is the hardest part and which is where nearly every attempt fails. If you take it back the first time something is dropped, you have taught both of you that the transfer was theatre.

Nadia handed Wes the entire school domain in September. In October he missed a form and their daughter did not get into the after-school program she wanted. Nadia knew about the form. She said nothing, which she describes as one of the more difficult things she has done, and which was the actual mechanism by which the domain moved.

What Symmetry Is Not

Two clarifications, because this argument can be misused in both directions.

Symmetry is not the goal. A perfectly symmetric division of every domain would be wildly inefficient and no functioning household does it. Specialization is good. The problem is never that a symmetry broke; it is that it broke by accident, in a direction nobody selected, and then became invisible.

The test is not whether the split is even. It is whether both of you could give the same account of why it is what it is, and whether that account survives being said out loud without either of you wincing.

And invariance is not rigidity. A rule that never changes regardless of circumstance is not a conservation law, it is a wall. What Noether requires is that the *laws* be constant, not that the state be frozen. A couple can renegotiate their arrangement every year and still be entirely invariant in the sense that matters, provided the process for renegotiating is itself stable and predictable.

The distinction is between changing the arrangement through a known procedure and changing it through mood. The first conserves trust. The second is the second household.

Practice: The Domain Map

Requires both. Ninety minutes, and it is more uncomfortable than it sounds.

Step one. Independently, each of you lists every recurring domain in your shared life. Not tasks; domains. Medical. School. Cars. Finances. Extended family on each side. Home repair. Food. Social calendar. Gifts. Insurance. Whatever else.

Step two. For each domain, each of you writes two names: who *executes* and who *remembers*. These are different and the difference is the whole exercise.

Step three. Compare. The disagreements about who remembers are where the invisible load lives, and there will be more of them than you expect, because remembering is by nature unobserved. The person who does not hold a domain generally has no idea that holding it is work, which is not a moral failure; it is a perceptual one, and this exercise is the correction.

Step four. Choose one domain to transfer completely, including the remembering. One. Not four. Set a review date six months out and agree, in advance and out loud, that the original owner will not intervene during that window.

What usually goes wrong: the transfer is made and then quietly reversed within eight weeks, through a series of small interventions that neither party registers as interventions. A reminder here, a "did you ever." Every one of those returns the model to its original owner. If you cannot tolerate the domain being run worse than you would run it, the transfer will not happen, and the honest move is to say that plainly rather than to perform an agreement you intend to undermine.

Emmy Noether

She was dismissed from Göttingen in 1933 under the Nazi civil service laws, moved to Bryn Mawr, and died two years later at fifty-three following surgery.

Her theorem is now taught in every graduate physics program in the world, and it says, in effect, that the things you can count on are produced by the things that do not change.

I have never encountered a better description of what it takes to be married to someone.

Chapter Fourteen: Phase Transitions, Or Why The Way Back Is Not The Way In

Put a pot of ice on a burner and watch the thermometer.

The temperature climbs steadily to zero degrees Celsius. Then it stops.

You are still adding heat. The burner has not changed. Energy is pouring into the pot at exactly the rate it was a minute ago, and the thermometer does not move. For several minutes, nothing appears to happen at all, and if you did not know better you would conclude the burner had failed.

What is happening during those minutes is that all the energy going in is being spent breaking the bonds of the crystal lattice. It is doing an enormous amount of work and producing no observable change in the quantity you are measuring. This is called latent heat, and for water it is large: melting a gram of ice takes roughly eighty times the energy required to raise a gram of liquid water by one degree.

Then the last of the ice goes, and the temperature starts climbing again.

I want to talk about what this explains, because I think it is the most quietly consoling fact in physics.

The Plateau

Every couple who has ever tried to repair something serious has run into the plateau.

They start doing the work. Therapy, or the practices in this book, or simply a sustained decision to behave differently. For the first stretch there is often rapid improvement, because the easiest problems go first and because attention alone produces an effect.

Then it stops. Weeks pass. Both people are working. Nothing changes. The conversations are better in form and identical in feeling. And the interpretation that arrives, in almost every case, is that the effort is not working and the relationship is not salvageable, because if this much effort produces nothing then the problem must be fundamental.

That interpretation is the single most common way that recoverable relationships end.

What the physics offers is an alternative reading. In a phase transition, there is a stage during which the entire energy input goes to restructuring rather than to moving the observable quantity. The system is not failing to respond. It is doing the expensive part, which by its nature does not show up on the instrument you are watching.

I want to be careful, because this can obviously be used to justify staying in anything forever, and I will draw the boundary in a moment. But the structural point stands: **in systems that change by phase transition rather than gradually, the absence of visible progress during a period of high input is expected and is not evidence of failure.**

The mistake is measuring temperature during melting.

What Actually Changes At A Transition

The deeper reason relationships behave this way is that they are not continuously adjustable systems. They are systems with distinct states.

Physicists describe a phase transition using something called an order parameter, a quantity that is zero in one phase and nonzero in the other. In a ferromagnet, it is the net magnetization. Above the critical temperature it is zero; below, it is not. The transition is not a matter of the material becoming gradually more magnetic. There is a threshold, and the qualitative character of the material changes across it.

Relationships have states of this kind, and everyone recognizes them without having a vocabulary for them. There is a state in which you assume good intent by default, and a state in which you do not. There is a state in which conflict is a problem to solve together and a state in which conflict is a contest. There is a state in which physical affection is unremarkable and a state in which it has become a signal that requires interpretation.

These are not points on a continuum. They are configurations, and the transitions between them are relatively fast and remarkably total. Couples describe crossing them in retrospect with startling precision: "somewhere around the second winter, something changed."

Which is why incremental effort often produces nothing until it produces everything. You are not sliding a dial. You are trying to change the phase.

Hysteresis

Now the part that I think is the most practically important idea in this book after decoherence, and the one that explains the most bitterness.

Take a piece of iron and put it in a magnetic field. As you increase the field, the iron magnetizes. Now reduce the field back to zero. The iron does not return to its original state; it stays magnetized. To get it back to zero magnetization you have to apply a field in the *opposite* direction, of substantial strength.

Plot the magnetization against the applied field and you get a loop rather than a line. The path out is not the path back. This is called hysteresis, from the Greek for "lagging behind," and it appears everywhere in nature: in magnets, in elastic materials, in ecosystems, in economies.

The defining feature is path dependence. **The state of the system depends not only on the current conditions but on the history of how it got there.**

Here is what that means for two people.

The conditions under which trust broke are not the conditions under which trust returns. If a relationship crossed into suspicion at a certain level of stress, reducing the stress back to that same level will not bring it back out. You have to overshoot: apply the reverse field, well past neutral, for a sustained period.

This is why "but I stopped doing the thing" is so consistently insufficient, and why the person who stopped experiences the continued distrust as unfair. From their position, the input has returned to its previous value, so the output should too. That is linear thinking, and the system is not linear. Restoring the prior conditions restores nothing, because the system remembers its path.

The relational translation is unwelcome and, I think, accurate: after a breach, behaving normally is not enough to return to normal. You have to behave better than normal, for longer than seems reasonable, and the requirement is not a punishment. It is what path dependence costs.

The person who understands this stops asking when they will be forgiven and starts running the reverse field. The person who does not understand it concludes that their partner is unwilling to move on, and that conclusion is the thing that usually finishes it.

Metastability

There is a third phenomenon here and it explains the endings that arrive without warning.

Water can be cooled below zero degrees without freezing, if it is very pure and undisturbed. This is called supercooling. The liquid state is metastable: it is not the lowest-energy configuration available, but there is a barrier to reaching the stable state, and without a nucleation site nothing initiates the change.

Then something perturbs it. A speck of dust. A tap on the glass. And the whole volume freezes almost instantly, propagating from the nucleation point.

An enormous number of relationships are supercooled. The stable state, given the current conditions, is ended. But the transition has a barrier, made of habit, logistics, children, mortgages, and the sheer effort of imagining anything else, and so the liquid state persists past the point at which it is stable.

And then something small happens. A remark at a party. A trip. An unrelated death that reorganizes someone's sense of time. And the transition that had been thermodynamically available for four years completes in six weeks, and everyone including both participants describes it as sudden.

It was not sudden. It was nucleated.

If you are the person left standing in the kitchen wondering how this happened so fast, the answer is that it did not. The conditions had been in place for a long time and the barrier was doing the work of making it look otherwise.

Critical Slowing Down

There is one more piece of this, and it is the only genuine early-warning indicator I know of.

Systems approaching a critical transition exhibit a characteristic warning sign. As the transition nears, the system's recovery from small disturbances becomes slower. Push it, and it takes longer to return to baseline than it used to. This is called critical slowing down, and it appears in models and in real data across an extraordinary range of systems: in ecosystems approaching collapse, in financial markets, in climate records preceding abrupt shifts.

It has been found in human psychology too. In 2014, Ingrid van de Leemput and a large team including Marten Scheffer published a paper in the Proceedings of the National Academy of Sciences reporting evidence of critical slowing down preceding transitions into and out of depressive episodes, using high-frequency emotional self-report data. Rising autocorrelation and variance in mood measurements preceded the shift.

I am not aware of a comparable study in couples, and I want to be honest that what follows is an extrapolation rather than a finding. But the indicator is worth watching, because it is measurable by anyone and it does not require any instrument.

How long does it take you to recover from a small conflict?

Not a large one. A small one, the kind that used to be over by dinner. If the recovery time has been growing, if a snapped remark that once cost twenty minutes now costs a day and a half, that is critical slowing down, and it is a signal about the state of the system rather than about the remark.

This is a far better early indicator than frequency of conflict, which is what most people watch. Frequency fluctuates for a hundred irrelevant reasons. Recovery time tracks something structural.

Nadia noticed, in year nine, that a disagreement about a dishwasher took four days to clear. In year three the same disagreement would not have survived the evening. Nothing about the dishwasher had changed.

Where This Frame Stops

The plateau argument is dangerous and I want to bound it.

"Keep working, the change is happening invisibly" is indistinguishable, from the inside, from "keep working, nothing is happening." Latent heat is a real phenomenon and so is a burner that has been switched off. The physics cannot tell you which one you are in.

Three things can help.

Set the duration in advance. Before you begin serious repair work, agree on a period, six months is common in the clinical literature, at the end of which you will assess honestly. A defined window converts an indefinite plateau into a bounded experiment.

Without one, the plateau argument extends forever and becomes a mechanism for avoidance.

Watch a different instrument. During melting, the temperature does not move, but other things do. You cannot see the lattice breaking, but you can see the ice getting smaller. Relationally, the equivalent measures are recovery time, the frequency of repair attempts being received rather than refused, and whether new subjects have become discussable. Those often move during a satisfaction plateau, and if none of them is moving after several months, that is meaningful.

And be honest about whether energy is going in. The plateau argument applies when both people are working. It does not apply when one person is working and the other is present. That is not a phase transition. That is a burner with one person standing next to it.

Practice: The Recovery Clock

This is a solo practice and it takes almost no effort.

For two months, note two things after any small conflict: the date, and roughly how long it took before you both felt normal again. A note on your phone is sufficient. Do not track the content. Do not track who was right. Only the duration.

At the end of two months, look at the trend rather than the values.

If recovery times are stable, whatever the absolute number, the system is in a steady state. It may be a state you like or do not like, but it is not currently in transition.

If they are lengthening, that is the signal, and the correct response is not to have fewer conflicts, which addresses nothing. It is to increase repair frequency, which is the intervention from Chapter Three, and to look at what has changed in the surrounding conditions over the same period.

If they are shortening, something you are doing is working, and you should find out what.

What usually goes wrong: people use the log as evidence in arguments. If the record becomes prosecutorial it stops being an instrument and becomes a weapon, and the measurement is no longer measuring the thing it was built for. Keep it to yourself, or agree explicitly that it is never cited in a fight.

The Consolation

I said at the top that latent heat is quietly consoling and I want to close on why.

Almost everyone I have talked to about a difficult stretch in a long relationship has described the same thing: a period during which they did everything right and nothing happened, and the nothing was worse than the difficulty had been, because the difficulty at least contained the possibility of a response.

What I would want them to know, and what the ice in the pot demonstrates every time anyone makes tea, is that there exists an entirely ordinary physical situation in which maximum energy input produces zero observable change, over a sustained period, right up until the moment the state changes completely.

The instrument is not broken. The energy is not being wasted. You are simply measuring the wrong quantity during the part where the structure is coming apart and reassembling.

And then, at some point that nobody can predict in advance, the last of it goes and the temperature starts climbing again.

PART FIVE

Integration: What To Do With All This

Chapter Fifteen: The Places It Breaks, And Why I Kept Going Anyway

Every metaphor has an edge. The useful thing about knowing where yours is, is that you can walk up to it and enjoy the view instead of falling off.

So here is the edge. It is shorter than you might expect, and I am not going to spend it apologizing.

The Scale Problem, Briefly

Quantum effects do not operate at the size of people.

The reason is decoherence, which we did in Chapter Three, and the numbers are not close enough to argue about. Max Tegmark ran the calculation in 2000 for the brain structures that had been proposed as sites of quantum computation and got decoherence times around ten to the minus thirteen seconds, against neural processes that run in milliseconds. That is roughly ten orders of magnitude of daylight.

Roger Penrose and Stuart Hameroff have disputed the calculation and are still working on it, and I want to be fair to them: they are serious people, quantum effects have turned up in biology in places nobody expected, including photosynthesis and the navigation systems of migrating birds, and the story is not finished. But it is a minority position, and nothing established says quantum coherence does any of your thinking for you.

For two whole human beings in a room, the case is closed. You are warm, enormous, and thoroughly entangled with the furniture.

So the bond between you and the person you love is not physically quantum.

I notice this bothers me far less than I expected, and I suspect it will bother you less too, because the thing we were after was never the mechanism. It was the shape. The shape is real, it turns up in both places, and the fact that a hundred-year-old argument about electrons keeps describing your marriage is a fact about how strange things arrange themselves, not a coincidence I need to explain away.

Where The Mappings Are Loose

A quick inventory, because you may have noticed some of these yourself and I would rather you know I did too.

The electron does not model the experimenter and adjust. Your partner does, constantly, and has been reverse-engineering you for years. That feedback loop has no counterpart in the physics at all, and it is arguably the most important thing about a long marriage.

Entangled particles are prepared once and that is the end of the story. No maintenance, no growth, no repair, no Thursday. Everything interesting about a relationship over time happens in territory the physics does not cover.

Complementarity is enforced by hardware. In your kitchen it is enforced by tiredness.

Tunneling has no agent. Nothing decides to try again. The deciding is the entire difficulty and it is completely absent from the equations.

Many worlds is contested, entropy is not quantum mechanics at all, and I put a chapter about Emmy Noether in a book with the wrong title on the cover because it was too good to leave out.

None of that is a confession. It is a map, and you can now read the rest of the book with it in your hand.

The Corrupted Forms

This is the part that matters, and it is the reason the chapter exists.

Every idea in here has an evil twin. They are all the same move: using a physics word to make somebody doubt something they correctly noticed, or to get out of something they owe.

Learn the twins. They are easy to spot once you know the shape, and spotting them is genuinely satisfying.

The observer effect, corrupted. *Reality is created by perception, so if you experience me as cruel, that is a fact about your observation.* This is gaslighting with a bibliography. The effect applies to interpretation, never to events. If it happened, it happened.

Uncertainty, corrupted. *You can never truly know another person, so you have no standing to hold me to anything.* Heisenberg was talking about conjugate variables. He was not talking about whether you came home when you said you would.

Superposition, corrupted. *I both want this and do not, and you have to accept the ambiguity.* Ambiguity about a feeling is legitimate and you should be given it. Ambiguity about conduct is not a state, it is a way of keeping somebody waiting, and "I contain multitudes" is not an answer to "are you staying."

Entanglement, corrupted. *We are bound beneath the level of choice, so you cannot leave.* This is the worst one. There is no bond anywhere in physics or anywhere else that overrides consent, and anyone reaching for cosmic vocabulary to explain why you may not go has told you something important about themselves.

Many worlds, corrupted. *Every choice happens somewhere, so none of this matters much.* The branches, if they exist, are sealed off and send nothing back. You are in this one.

Coherence, corrupted. *We should want the same things, so your different want is a failure of alignment.* Coherence is a stable relationship between differences. A demand for identical phase is a demand for somebody to disappear.

If you catch one of these being run on you, by a partner or by an author, the correct inference is that the plain-language version would not have survived being said plainly.

One Thing I Will Not Be Light About

The frameworks in this book do not apply to abuse, and the language of quantum romance is one of the tools by which abusive relationships are sustained. The observer effect becomes an argument that you are misperceiving. Entanglement becomes an argument that you cannot leave. Tunneling becomes an argument to keep trying.

There is no interpretation under which fear of your partner is a complementary perspective.

If that describes your situation, the book you need is not this one. In the United States, the National Domestic Violence Hotline is 1-800-799-7233, and comparable services exist nearly everywhere and can be found in about a minute.

The Softness Of My Other Field

The relationship science I have cited is real, and it is softer than physics, and you should hold it more loosely than the parts about electrons. Effect sizes are modest. Samples skew Western and educated. Psychology has had a rough decade with replication, and some of what I have quoted has never been through a serious replication attempt.

The most honest study in the field is probably Samantha Joel's 2020 machine-learning analysis pooling data from more than eleven thousand couples, which found that the best predictor of how a relationship goes is how the person inside it already feels about it, that partner characteristics predict almost nothing, and that even the best models leave most of the variance unexplained.

Which is to say: nobody knows. Not the researchers, not the therapists, not me.

I find that wonderful rather than depressing. We have mapped the interior of the proton and we cannot tell you whether two people will be happy in four years.

Why I Kept Going

Because the shapes are real even where the mechanism is not, and because human beings think in shapes.

That measurement is never passive. That some questions exclude each other. That private worlds erode through contact and have to be defended on purpose. That a pair can be fully specified while its members are not. That certain goods trade off by nature and can only be distributed, never escaped. That ambiguity is usually a property of the question. That thinning a wall beats charging it. That order has a running cost.

Every one of those is true in plain language with no physics attached, and almost nobody arrives at them unaided. I did not. I got them from a long argument between people who were trying to understand something completely different and ended up describing my life.

That is what a good metaphor does. It hands you a shape from somewhere else, and you recognize your own situation inside it, and you are never able to unsee it again.

The scaffolding comes down eventually. The building stays up.

Chapter Sixteen: The Twelve Week Practice

This chapter contains no physics.

That is deliberate. If the ideas in this book are any good, they should be statable and practicable without the vocabulary, and a program that requires you to keep thinking about photons in order to work is a program that will stop working the week you get busy.

Twelve weeks. One practice per week, each introduced on top of the last, most of them small. By week twelve you should be running about forty minutes of deliberate maintenance a week plus a handful of habits that cost nothing.

A few notes before you start.

This is designed to be done by one person if necessary. Roughly half the practices require a partner's participation and half do not. If your partner is unwilling, or skeptical, or simply busy, start with the solo ones. Several of them change the dynamic on their own, and a partner who has watched something improve for a month is a great deal easier to recruit than one who has been handed a book.

Do not start all twelve at once. The reason this is sequenced is that each week's practice is small enough to survive a bad week. Twelve practices adopted simultaneously is a New Year's resolution, and you know what happens to those.

Expect week five or six to be where it falls apart. That is the normal attrition point. When you miss one, resume the next day rather than restarting the program, because restarting is a form of quitting that feels like diligence.

Week One: The Two Columns

Solo practice, from Chapter One.

Once this week, after a moment when you have concluded something about your partner's inner state, take three minutes and write two columns. Left: what a camera would have recorded. Right: what you concluded. Then generate three alternative explanations consistent with the same camera data.

That is the entire week. You are not asked to change any behavior. You are building the capacity to notice the seam between observation and interpretation, and that noticing is the foundation for everything else.

By Sunday you should have done this two or three times. If you did it once, that is fine.

Week Two: Naming The Arrangement

Requires both. From Chapter Two.

Adopt three words as shared vocabulary. **Witness** means listen and do nothing. **Counsel** means give me your actual opinion. **Company** means nothing needs discussing, I want to not be alone.

The practice is to name one before any conversation with emotional weight, and to ask if the other person has not named one. Four seconds. Nothing else changes.

The failure mode is that the habitual problem-solver will slip. When they do, the correction is one word, said flatly: "witness." Not a complaint. An instrument setting.

Keep this one forever. It is the highest return-per-effort practice in the book.

Week Three: The Enclosure

Requires both. From Chapter Three.

Twenty-five minutes, twice this week, on specific days. Put them on a calendar now, not in your head. Monday and Thursday work well for most people.

Devices in a different room. Not silenced, not face down. In a different room.

Three prohibited topics: scheduling, money, and anything with a task attached. Those subjects are not banned from your life, only from this window.

What goes in instead is recounting. Things that do not need to be said. A thing you noticed. A memory that surfaced. A thought with no conclusion.

The first two sessions will contain long silences and will feel artificial. That is expected and it passes somewhere around session four.

Keep this one forever as well. If you only do two practices from this book, make them this one and week two.

Week Four: The Raw Channel

Requires both, and works at any distance. From Chapter Five.

Twice a day, send something unprocessed. The standard is that it should feel slightly not worth sending. A photograph of nothing. Three words. A thought that went nowhere.

The rule that makes it work: no response required. This is broadcast, not correspondence. If you find yourself keeping score of who sent more, say so out loud immediately, which is week one's practice arriving in the wild.

Week Five: Turning Toward

Solo practice, and it is the hardest one.

For one week, notice small bids for your attention and respond to them. Your partner comments on something outside the window. Reads something aloud. Puts a hand on your shoulder while you are working.

The practice is to stop what you are doing, for three to five seconds, and respond. Not with enthusiasm you do not have. With acknowledgment.

You will discover two things. First, that there are far more of these than you knew. Second, that you have been declining most of them for years without any awareness of doing so, because declining is silent.

This is the week most people find genuinely uncomfortable, and it produces more change than any other single item here.

Week Six: The Rotation

Solo practice, from Chapter Seven. One sitting, about an hour.

Take an internal contradiction you have been carrying. Write it in binary form. Then write the question you have been asking, which will usually contain the word *really*.

Then generate three genuinely different questions the same material could be an answer to, and write a paragraph answering each. Look for the one you write quickly and without qualification.

This will not produce a decision. It produces a better question, which is a precondition for one.

Week Seven: The Break Protocol

Requires both. From Chapter Eight.

Agree on two things in advance, while nothing is wrong.

First, either of you may call a break during a conflict, and the person calling it must state a return time. Forty minutes is a good default. The stated return is the entire content of the agreement; a break without one reads as abandonment and makes everything worse.

Second, the person who wants to keep going agrees to stop when a break is called, without a final point.

You will not be able to do this the first time. Almost nobody can. Agree that failing at it is expected and that you will try again.

Week Eight: The Configuration Audit

Requires both. From Chapter Nine. One sitting, about forty minutes, plus two calendar entries.

List everything substantial you have done together in the past three months. Sort into familiar-and-safe versus novel-or-arousing. Observe the ratio, which will surprise you.

Then book two things for the next quarter that genuinely qualify. The test: neither of you has done it, you might be bad at it, other people will be present, and imagining it produces some reluctance. If it sounds nice, it does not qualify.

Pay for them if paying makes attendance more likely. It does.

Week Nine: The Eleven Words

Solo practice, from Chapter Ten. Possibly the most consequential week here.

Identify something you have not said and would like to have said. Not the hardest thing. Something in the middle.

Write the full version. Then cut it to one sentence, under fifteen words, containing one true thing, no accusation, and no request.

Then wait for a low-barrier moment. A car, a walk, a kitchen at night, anything with parallel attention and no eye contact. Say it. Add nothing. Let the silence happen.

Give yourself two weeks for the moment to arrive. If it has not, manufacture one. Suggest a drive.

Week Ten: The Phase Audit

Requires both. From Chapter Six. Ninety minutes, scheduled when nothing is wrong.

List the ten or twelve subjects that most affect your shared life. Independently, each of you writes where you stand and where you believe the other stands.

Compare. The gaps between what you believed and what is true are the finding, and there will be at least one that is several years stale.

Then sort each subject: aligned, stably different, or requires resolution. Anything in the third category gets a date on a calendar before you leave the room.

Resist putting everything in the second category. It is comfortable and it is where avoidance goes to live.

Week Eleven: Closing A Branch

Solo practice, from Chapter Eleven. One hour, written.

Take a counterfactual you return to. Write what you imagine. Then populate it honestly with everything your imagination omits: the money, the fight you would be having by year three, what happened to the people you would not have been there for, what is dull about that life.

Then write the decision you actually made, in the first person, active voice, as a decision rather than as a thing that happened to you.

Then one sentence about something specific this life has that the other one does not.

Week Twelve: The Maintenance Schedule

Requires both. From Chapter Twelve. One evening.

List the components of your shared system: accurate models of each other, division of labor, repair process, shared account of your history, unstructured contact, physical intimacy, alignment on the big questions, inputs from outside.

For each, note when it was last attended to on purpose. Note what is drifting. Then assign a service interval.

Then put the recurring items on a real calendar, with alerts, and make sure the invitations are not all coming from one person. If one of you becomes the maintenance department, the audit has failed regardless of what the calendar says.

Two Optional Additions

Two practices from later chapters did not fit the weekly sequence, because both operate on a longer timescale. Add them when the twelve weeks are done.

The Domain Map, from Chapter Thirteen, is a ninety-minute audit of who executes and who remembers in every recurring area of your shared life, followed by the complete transfer of exactly one domain, remembering included, with a six-month review date and an advance agreement that the original owner will not intervene. It is the most reliable way I know to move an unchosen division of labor, and it fails in exactly one way, which is that the transfer gets quietly reversed through small reminders that nobody registers as reminders.

The Recovery Clock, from Chapter Fourteen, is a solo two-month log of how long it takes the two of you to feel normal again after small conflicts. Not the content, not who was right, only the duration. The trend is the finding. Lengthening recovery is the earliest structural warning signal available to you, and it is visible long before anything else is.

What To Keep

At the end of twelve weeks, most people cannot sustain all of it, and should not try.

If you keep three things, keep the enclosure from week three, the arrangement vocabulary from week two, and the break protocol from week seven. Those three are cheap, they run indefinitely, and between them they cover maintenance, misfired conversations, and escalation, which is most of what goes wrong on an ordinary week.

If you keep one thing, keep the enclosure. Twenty-five minutes, twice a week, no devices, no logistics. It is the smallest reliable dose of the thing that everything else in this book is ultimately about, which is two people paying uninterrupted attention to each other on purpose, at a time they agreed on in advance, for no productive reason at all.

A Note On Doing This Alone

Some of you are reading this without a partner, or with one who will not participate.

Weeks one, five, six, nine, and eleven are entirely solo and are, in my view, the ones with the deepest effects, because they change what you perceive rather than what you negotiate. A person who has stopped confusing observation with interpretation, who responds to bids, who can rotate an internal contradiction, who can say a true thing in eleven words, and who has taken authorship of an old decision is a substantially different person to be in a relationship with.

That is not a consolation prize. It is most of the mechanism.

Conclusion: What The Strangeness Was For

Here is what I think happened to us.

Somewhere in the last two hundred years we agreed, without anyone signing anything, that love was a solved problem. Mysterious in its arrival, perhaps, but structurally simple once it got here. Two people, definite properties, a fact of the matter about whether it was working. Any confusion you experienced was a personal failing or a sign you had picked wrong.

And then we all went out and lived actual lives, in which nothing behaved like that at all.

The physicists had the same century and handled it better. When their world stopped making sense they did not conclude that the world was defective. They concluded that their categories had run out, and they went and built new ones, and it took twenty-five years and several nervous breakdowns and it worked. Everything you own runs on the vocabulary they invented in a state of considerable distress.

That is the whole argument of this book, and the metaphors were only ever the delivery mechanism.

When a thing you care about stops making sense, there are two available responses. You can decide the thing is broken. Or you can consider that your categories have run out and go looking for better ones.

Almost everybody chooses the first. Almost every relationship that ends badly ends with at least one person convinced they were dealing with a defective situation, when what they had was an ordinary situation described in a vocabulary too crude to hold it.

Nadia thought she was married to a man who did not want to be there. She was married to a man with a sick mother and a decade of learned caution, and her instrument could not detect either.

Priya thought she had lost something. She had a calendar.

June thought her grief was malfunctioning because it kept returning two answers. It was returning one answer to a question she had not thought to ask.

Dev thought he had a life in Lisbon he was failing to live. He had a piece of fiction he wrote when he was thirty-one and frightened.

None of them needed physics. They needed a bigger set of shapes to think with, and physics is simply where the best shapes are kept, because physics is the field that has been forced most often to give up what it was certain about.

What I Actually Believe

Here, without hedging, is what I think is true about love, arrived at through several years of reading physics I only partly understand and psychology I understand better and living a life that has taught me more than either.

Attention is the whole thing. Not feeling. Attention. Every practice in this book is a mechanism for producing more of it or wasting less of it, and if the practices vanished tomorrow and you simply gave someone forty uninterrupted minutes a week, you would get most of the benefit.

Most relationship pain is a description problem. Not all. Some of it is a real problem with a real cause and a necessary decision at the end of it, and I have tried not to pretend otherwise. But a startling proportion of ordinary suffering between two decent people is generated by asking a question that has no answer and then treating the absence of an answer as evidence of a catastrophe.

Erosion is the default and requires no villain. This is the single idea I would most want a young person to have, because it arrives too late for almost everyone. Nothing has to go wrong for a relationship to end. It only has to be left alone.

Certainty about another person is not available and is not the goal. You will not find out what your partner really thinks in some final way. Nobody has ever done this. What is available is a long, revisable, continuously updated model, maintained through contact, which is a much better thing than certainty and which people abandon constantly because it is not what they were promised.

And it is voluntary. Every day, both of you, choosing again. There is no bond that removes that, and any framework that tells you otherwise, including one built out of quantum mechanics by someone who means well, is doing you harm.

The Last Thing

At the end of his life Niels Bohr was asked what the opposite of a profound truth was, and he is supposed to have said that the opposite of a small truth is a falsehood, but the opposite of a great truth is another great truth.

I have never been able to verify the quotation, which is fitting.

But I have thought about it for years, and I have come to believe it is the most useful sentence ever produced by a physicist about anything other than physics. It is the whole content of complementarity, delivered in one line, and it is what I would leave you with.

You want to be known and you want to be free. You love someone and you are sometimes unbearably tired of them. You would not change your life and you think about the one you did not take. He is the person you trust most in the world and he has hurt you in ways he does not know about.

None of those pairs is a contradiction to be resolved. They are the shape of the actual thing, and the shape is strange, and the strangeness is not a sign that you have done it wrong.

It is a sign that you are looking at something real.

Appendix One: A Glossary Of The Physics

Amplitude. The quantity a quantum state assigns to each possible outcome. Amplitudes are complex numbers, which is why they can cancel one another, which is why interference exists. Squaring an amplitude gives a probability.

Bell's theorem. John Stewart Bell's 1964 proof that any theory in which distant systems have predetermined properties and no influence travels faster than light must obey a specific mathematical limit on correlations. Quantum mechanics predicts violations of that limit. Experiments confirm the violations.

Coherence. A stable phase relationship between waves. Coherent sources produce stable interference patterns; incoherent ones do not. Coherence does not mean sameness. It means the relationship between the components holds still.

Complementarity. Bohr's principle that certain descriptions of a system are each valid, each necessary, and mutually exclusive in application, because the experimental arrangements required for each exclude one another.

Conjugate variables. Pairs of properties, such as position and momentum, related mathematically such that sharpness in one entails spread in the other. The relationship is a property of waves rather than a limitation of measurement.

Decoherence. The process by which a quantum system loses its interference behavior through interaction with its environment, as information about the system's state leaks outward. Extremely fast for anything large or warm. The reason the everyday world looks classical.

Entanglement. A condition in which the joint state of two or more systems is fully specified while the individual states are not.

Produces correlations stronger than any classical account permits. Cannot be used to transmit information.

Interference. The addition and cancellation of wave amplitudes, producing patterns of reinforcement and suppression. The signature phenomenon of quantum behavior.

Many-worlds interpretation. Hugh Everett's 1957 proposal that the wave function never collapses and that all measurement outcomes occur, in branches that decohere from one another. One interpretation among several; not established.

Measurement problem. The unresolved question of why quantum mechanics describes spreads of possibility while experience delivers single outcomes. Decoherence explains part of it and not all of it.

No-communication theorem. The proof that entanglement cannot be used to send a signal. Each party's local results are random regardless of what the other does; the correlation is only visible when the data are compared through ordinary channels.

Nonlocality. The feature of quantum correlations that they do not weaken with distance and cannot be explained by any local mechanism. Not a signal, not a force, and not a channel.

Phase. Where a wave is in its cycle at a given moment. Differences in phase determine whether waves reinforce or cancel.

Second law of thermodynamics. The principle that the entropy of an isolated system does not decrease. Local order can be maintained only by continuous energy input and by exporting entropy elsewhere. Not quantum mechanics; included in this book anyway.

Squeezing. Deliberately reducing uncertainty in one conjugate variable at the cost of increasing it in the other, keeping the product at its minimum. Used in gravitational wave detection.

Superposition. A quantum state expressed as a combination of other states. Not a mixture of definite conditions we happen to be ignorant of. A superposition with respect to one measurement basis is a definite state with respect to another.

Tunneling. The passage of a quantum system through an energy barrier it lacks the energy to surmount, arising because the wave function extends into and beyond the barrier. Probability falls exponentially with barrier width. Responsible for stellar fusion and for radioactive decay.

Wave function. The mathematical object describing a quantum system, assigning amplitudes to possible outcomes and evolving smoothly according to the Schrödinger equation.

Wave-particle duality. The historical name for the fact that quantum systems display wave-like or particle-like behavior depending on the experiment. Modern physics regards this as one entity described by two inadequate classical concepts, not as two natures.

Appendix Two: A Glossary Of The Relationship Science

Affect labeling. Putting emotional experience into words. Associated with reduced amygdala reactivity and improved regulation.

Attachment. The behavioral system organizing proximity-seeking toward a caregiver or partner, described by John Bowlby and extended to adult romantic relationships by Cindy Hazan and Phillip Shaver.

Bid for connection. A small, low-stakes attempt to engage a partner's attention. Gottman's term. The response to bids, aggregated over thousands of instances, is a better index of relationship health than the handling of major events.

Coregulation. The process by which partners stabilize one another's physiological and emotional states, such that each person's baseline comes to depend on the other's availability.

Demand-withdraw pattern. A self-amplifying cycle in which one partner pursues discussion and the other avoids it, each responding to the other. One of the most robust predictors of relationship distress.

Dependency paradox. Brooke Feeney's finding that partners who receive more responsive support subsequently show more independent functioning, not less.

Destiny and growth beliefs. Implicit theories of relationships, studied by Raymond Knee. Destiny believers treat difficulty as evidence of mismatch; growth believers treat it as a problem to work on.

Emotional granularity. The precision with which a person differentiates their emotional states. Higher granularity is associated with better regulation.

Flooding. Physiological overwhelm during conflict, marked by elevated heart rate, after which productive processing largely stops. Recovery takes roughly twenty minutes.

Mental load. The anticipating, monitoring, and deciding components of household labor, distinct from execution and unevenly distributed.

Michelangelo phenomenon. Caryl Rusbult's finding that partners who perceive and behave toward you in ways aligned with your own ideal self help move you toward it.

Positive illusions. Idealized perceptions of a partner, associated with greater satisfaction and stability, and with partners drifting toward the idealized version over time.

Repair attempt. Any move that interrupts an escalating conflict. Whether repair attempts are received matters more than whether they are skillful.

Responsiveness. The perception that a partner understands you, values you, and cares. The core of the intimacy process described by Reis and Shaver.

Self-expansion. Arthur Aron's model in which relationships are motivated partly by the incorporation of another's perspectives and resources into the self. Novel, arousing shared activity increases relationship quality.

Shared reality. The mutual confirmation of interpretations of the world, associated with closeness and meaning.

Sliding versus deciding. Stanley, Rhoades, and Markman's distinction between relationship transitions entered deliberately and those arrived at by accumulated circumstance.

Sudden gains. Discrete, large improvements between therapy sessions, associated with better outcomes.

Appendix Three: Notes And Further Reading

This section gives sources by chapter, in plain prose. Where I have described research, you can find it here. Where a claim has no source listed, it is my own observation and should be weighted accordingly.

On The Physics Generally

For a serious but readable treatment of quantum foundations, David Lindley's *Where Does the Weirdness Go?* and Adam Becker's *What Is Real?* are both excellent, the latter being especially good on the history of the interpretation disputes. Sean Carroll's *Something Deeply Hidden* argues for many worlds with unusual clarity about what is and is not established. For decoherence specifically, Wojciech Zurek's 1991 article in Physics Today, "Decoherence and the Transition from Quantum to Classical," remains the clearest short account, building on H. Dieter Zeh's 1970 paper in Foundations of Physics.

Chapter One

Thomas Young's double slit work dates to 1801. Werner Heisenberg's uncertainty paper appeared in 1927; the microscope argument is in it, and he later moved away from it as the explanation of the principle.

The behavioral confirmation study is Snyder, Tanke, and Berscheid, "Social Perception and Interpersonal Behavior," Journal of Personality and Social Psychology, 1977. The attribution literature is best entered through Bradbury and Fincham, "Attributions in Marriage: Review and Critique," Psychological Bulletin, 1990. The positive illusions work is Murray, Holmes, and Griffin, published in two 1996 papers in the Journal of Personality and Social Psychology, the second of which is subtitled "The Self-Fulfilling Nature of Positive Illusions." Rosenthal and Jacobson's *Pygmalion in the Classroom* appeared in 1968; treat its effect sizes with caution.

Chapter Two

Bohr introduced complementarity at Como in September 1927. The Reis and Shaver intimacy process model is in "Intimacy as an Interpersonal Process," a 1988 chapter in the Handbook of Personal Relationships. On desire and security, Esther Perel's *Mating in Captivity* (2006) is the accessible source; it is clinical observation rather than controlled research and should be read as such.

Chapter Three

Peter Berger and Hansfried Kellner's "Marriage and the Construction of Reality" appeared in Diogenes in 1964 and is short enough to read in a sitting. On repair and the ratio findings, John Gottman's *The Seven Principles for Making Marriage Work* is the popular treatment; for the methodological critique, see Heyman and Slep, "The Hazards of Predicting Divorce Without Crossvalidation," Journal of Marriage and Family, 2001.

Chapter Four

The EPR paper is Einstein, Podolsky, and Rosen, Physical Review, 1935. Schrödinger's response, in which he coined "entanglement," appeared in the Proceedings of the Cambridge Philosophical Society the same year. Bell's paper is "On the Einstein Podolsky Rosen Paradox," Physics, 1964. Key experiments: Freedman and Clauser 1972; Aspect, Dalibard, and Roger 1982; and the loophole-free tests of 2015, of which Hensen and colleagues in Nature is the most cited.

The monogamy result is Coffman, Kundu, and Wootters, "Distributed Entanglement," Physical Review A, 2000.

On the psychology: Levenson and Gottman, "Marital Interaction: Physiological Linkage and Affective Exchange," JPSP, 1983. Sbarra and Hazan, "Coregulation, Dysregulation, Self-Regulation," Personality and Social Psychology Review, 2008. Coan, Schaefer, and Davidson, "Lending a Hand," Psychological Science, 2006. Ruth Feldman's synchrony work is spread across many papers; her 2012 review in Parenting: Science and Practice is a reasonable entry point. On differentiation, Murray Bowen's *Family Therapy in Clinical Practice.*

Chapter Five

Long-distance entanglement demonstrations include the 143-kilometer experiment between La Palma and Tenerife reported by Zeilinger's group in 2012 and the 1,200-kilometer satellite experiment reported by Pan Jianwei's group in Science in 2017.

Laura Stafford's *Maintaining Long-Distance and Cross-Residential Relationships* (2005) is the standard review. Jiang and Hancock, "Absence Makes the Communication Grow Fonder," Journal of Communication, 2013. On post-reunion dissolution, Stafford, Merolla, and Castle, Journal of Social and Personal Relationships, 2006.

Chapter Six

On shared reality, Hardin and Higgins's 1996 chapter, and Rossignac-Milon, Bolger, Zee, Boothby, and Higgins, "Merged Minds: Generalized Shared Reality in Dyadic Relationships," JPSP, 2021. The Michelangelo phenomenon is summarized in Rusbult, Finkel, and Kumashiro, Current Directions in Psychological Science, 2009. On cognitive labor, Allison Daminger, "The Cognitive Dimension of Household Labor," American Sociological Review, 2019.

Chapter Seven

On emotional granularity, Kashdan, Barrett, and McKnight, Current Directions, 2015, and Lisa Feldman Barrett's *How Emotions Are Made* for the broader theory. Affect labeling: Lieberman and colleagues, "Putting Feelings Into Words," Psychological Science, 2007. On age and emotional complexity, Carstensen, Pasupathi, Mayr, and Nesselroade, JPSP, 2000. On culture and mixed emotions, work by Miyamoto, Uchida, and Ellsworth. On ambivalent ties and health, Bert Uchino's program of research, summarized in his 2004 book *Social Support and Physical Health*. On the quantum cognition program, Busemeyer and Bruza, *Quantum Models of Cognition and Decision* (2012), which is explicitly a formalism and not a claim about brain physics.

Chapter Eight

Heisenberg 1927 for the original. On measurement-disturbance relations distinct from the conjugate-variable relation, Masanao Ozawa's work from 2003 onward. Relational dialectics: Baxter and Montgomery, *Relating: Dialogues and Dialectics* (1996). The dependency paradox: Brooke Feeney, "The Dependency Paradox in Close Relationships," JPSP, 2007. Demand-withdraw: Christensen and Heavey, JPSP, 1990, and much subsequent work.

Chapter Nine

Einstein's photoelectric paper is from 1905; de Broglie's matter wave hypothesis from 1924. The quantitative duality relation is Englert, "Fringe Visibility and Which-Way Information," Physical Review Letters, 1996.

On love's components: Acevedo and Aron, "Does a Long-Term Relationship Kill Romantic Love?", Review of General Psychology, 2009, and Acevedo, Aron, Fisher, and Brown, "Neural Correlates of Long-Term Intense Romantic Love," Social Cognitive and Affective

Neuroscience, 2012. On self-expansion and novelty, Aron, Norman, Aron, McKenna, and Heyman, JPSP, 2000. The suspension bridge study is Dutton and Aron, JPSP, 1974.

Interlude

The quantum Zeno effect is Misra and Sudarshan, Journal of Mathematical Physics, 1977, and has since been demonstrated experimentally. The no-cloning theorem is Wootters and Zurek, Nature, 1982, with Dieks arriving independently the same year. Bell's sock story opens "Bertlmann's Socks and the Nature of Reality," Journal de Physique, 1981, and is the most charming paper ever written about a serious theorem. The Casimir effect was predicted in 1948 and first measured convincingly by Lamoreaux in 1997. Schrödinger's cat appears in his 1935 paper "Die gegenwärtige Situation in der Quantenmechanik," where it is unambiguously a joke at the theory's expense.

Chapter Ten

George Gamow's tunneling account of alpha decay dates to 1928. On bids and turning toward, Gottman's popular works; on the critique, Heyman and Slep as above. Sudden gains: Tang and DeRubeis, Journal of Consulting and Clinical Psychology, 1999, with substantial subsequent replication and debate.

Chapter Eleven

Everett's thesis was published in condensed form in Reviews of Modern Physics in 1957. For the modern development, David Wallace's *The Emergent Multiverse* (2012). For alternatives: de Broglie-Bohm theory, and the objective collapse model of Ghirardi, Rimini, and Weber (1986), which is experimentally testable in principle.

On affective forecasting, Gilbert, Pinel, Wilson, Blumberg, and Wheatley, "Immune Neglect," JPSP, 1998, and Daniel Gilbert's *Stumbling on Happiness*. The Olympic study is Medvec, Madey, and Gilovich, JPSP, 1995. Maximizing: Schwartz and colleagues, JPSP, 2002, and Barry Schwartz's *The Paradox of Choice*. Implicit theories: Knee, JPSP, 1998. Sliding versus deciding: Stanley, Rhoades, and Markman, Family Relations, 2006.

Chapter Twelve

Baumeister, Bratslavsky, Finkenauer, and Vohs, "Bad Is Stronger Than Good," Review of General Psychology, 2001. Ilya Prigogine's work on dissipative structures earned the 1977 Nobel Prize in Chemistry; *Order Out of Chaos*, with Isabelle Stengers, is the popular treatment. Eli Finkel, *The All-or-Nothing Marriage* (2017).

Chapter Thirteen

Emmy Noether's theorem was published in 1918 as "Invariante Variationsprobleme." For an accessible account of both the mathematics and the life, Dave Goldberg's *The Universe in the Rearview Mirror* and Ransom Stephens's *The Left Hand of the Electron* both treat it well; Leon Lederman and Christopher Hill's *Symmetry and the Beautiful Universe* is the fuller popular treatment.

On equity theory, the foundational work is Walster, Berscheid, and Walster in the 1970s, developed extensively by Elaine Hatfield. On cognitive labor and how unequal arrangements come to feel chosen, Allison Daminger, "The Cognitive Dimension of Household Labor," American Sociological Review, 2019, and her subsequent work on how couples explain their divisions.

Chapter Fourteen

Latent heat and phase transitions are standard thermodynamics; any introductory text covers them. Hysteresis in ferromagnetic materials is likewise textbook, and the concept has been productively exported to ecology, where Marten Scheffer's *Critical Transitions in Nature and Society* (2009) is the standard reference.

On critical slowing down as an early warning signal, Scheffer and colleagues published a widely cited overview in Nature in 2009, "Early-Warning Signals for Critical Transitions." The application to human mood is van de Leemput and colleagues, "Critical Slowing Down as Early Warning for the Onset and Termination of Depression," Proceedings of the National Academy of Sciences, 2014. I am not aware of an equivalent study in couples, and my extension of the indicator to conflict recovery time is an extrapolation rather than a finding.

Chapter Fifteen

Max Tegmark, "Importance of Quantum Decoherence in Brain Processes," Physical Review E, 2000. The Penrose-Hameroff orchestrated objective reduction proposal is set out in a number of places; their 2014 review in Physics of Life Reviews includes responses to critics. Fritjof Capra's *The Tao of Physics* (1975) and Gary Zukav's *The Dancing Wu Li Masters* (1979) are the origin points of the popular genre. Alan Sokal's hoax paper appeared in Social Text in 1996.

On the state of relationship prediction, Samantha Joel and a large collaborative team, "Machine Learning Uncovers the Most Robust Self-Report Predictors of Relationship Quality Across 43 Longitudinal Couples Studies," Proceedings of the National Academy of Sciences, 2020. It is the most important paper in the field in a decade and it is humbling.

About The Author

Erica Fargo writes about the places where science and ordinary life intersect. She has spent the better part of a decade reading physics she was not trained in and psychology she was, and this book is the result of trying to reconcile the two without lying about either.

She lives on the coast, keeps an unreasonable number of unfinished notebooks, and remains unable to explain the measurement problem at a dinner party without losing the room.

www.ingramcontent.com/pod-product-compliance
Lightning Source LLC
LaVergne TN
LVHW050818061126
851857LV00012B/871